Blueprint Reading
Made Easy

Stanley H. Aglow

BNP Business News Publishing Company
Troy, Michigan

Editor: Joanna Turpin
Production Coordinator: Mark Leibold

Library of Congress Cataloging in Publication Data

Aglow, Stanley H., 1927 -
    Blueprint reading made easy/Stanley H. Aglow
       p. cm.
    Includes index.
    ISBN 0-912524-71-5
    1. Blueprint   I. Title.
T379.A35   1992
604.2'5--dc20   92-24333
                                CIP

Cover Photograph: SuperStock, Inc.

Printed in United States
7 6 5 4 3 2 1

DISCLAIMER

This book is only considered to be a general guide. The author
and publisher have neither liability nor can they be responsible to
any person or entity for any misunderstanding, misuse or
misapplication that would cause loss or damage of any kind,
including material or personal injury, or alleged to be caused
directly or indirectly by the information contained in this book.

ACKNOWLEDGEMENTS

For their help in supplying the necessary blueprints and drawings,
I wish to thank the following companies:

The Kling-Lindquist Partnership, Inc.; Architects and Planners of
Philadelphia, PA; Houston Instrument, Austin, TX; Hewlett-
Packard Company, San Diego, CA; Hart & Cooley Mfg.,
Holland, MI; Bryant Air Conditioning Mfg., Indianapolis, IN;
Ruud Air Conditioning Mfg., Fort Smith, AR; Vaillant Mfg.
Cinnaminson, NJ; The Hydronics Institute, Berkeley Heights, NJ;
National Concrete Masonry Assoc., Herndon, VA; Rheem Mfg.
Company; Northwest Blueprint and Supply Company.

TABLE OF CONTENTS

Chapter 8

Construction specifications are the written guidelines outlining the work and duties of the engineer, architect, contractor, and owner. These specifications and the working blueprints are normally included in the legal contracts signed by all parties.

Chapter 9

Schedules provide detailed information about particular items on blueprints. This information is presented in the form of a table, with reference numbers corresponding to numbers on the blueprint.

Assignment 1

The first assignment illustrates some of the information a contractor must know before proceeding with the construction of a house.

Assignment 2

This assignment illustrates some of the information necessary when constructing an office building.

Assignment 3

HVAC contractors must understand the piping associated with hydronic systems. This assignment focuses on the layout of this piping.

Assignment 4

The kitchen plan is but a small part of a total building plan. The blueprints in this assignment cover some of the information required when constructing a kitchen.

INTRODUCTION

When designing a building, the architect and engineer must create explicit drawings detailing every aspect of the structure. These drawings include specific information concerning structural aspects, as well as plumbing, electricity, heating, ventilating and air conditioning. The many contractors who are hired to construct the building must then interpret these drawings correctly to ensure accurate construction takes place. These drawings are called *blueprints*.

Service technicians and installation mechanics in the HVAC industry must be able to read and interpret blueprints, then take this information and successfully install and service HVAC systems. Many service technicians find the lines and symbols on a blueprint confusing, as many technicians have no formal training in blueprint reading. This book will help eliminate this confusion.

It is important that the service technician understand the relationship between the HVAC equipment and the highly technical building construction; a heating, cooling and ventilation system is not a separate and independent part of a building, but an integral element of the total construction. This is a significant concept, because the energy used to operate a heating, cooling, and ventilation system is the greatest expense in large building operation. The service technician plays an important role in determining the energy savings of a building through proper installation and planning. Blueprints facilitate this proper installation and planning.

The objective of this book is to give the reader information concerning construction blueprints; that is, how to read these blueprints, and how these blueprints relate to one another. This book will also discuss the relationship and responsibility of each of the following: (1) the building owner who supplies the financing and reason for construction, (2) the architect who creates a drawing of the building, (3) the engineers who work under the direction of the architect and detail all construction requirements to ensure a safe building, and (4) the building contractors who take all the information and transform it into a building. It is the building contractor who has the most important job of all; reading the blueprints and ensuring good, solid construction takes place.

This book contains a series of drawings and explanations, which are designed to eliminate the confusion often encountered when reading blueprints. **It is important to note that some of the drawings contained in this book are designed for instructional use only and may vary from actual blueprints.**

If this book is not being used in conjunction with an instructor, the following suggestions will be of great help:

1. Read one chapter at a time and make notes concerning items that are confusing. Then, go back and re-read the sections that are unclear.

2. Read the chapter again and this time use the sample drawings that are referred to in the text.

3. Answer the questions at the end of the chapter before proceeding to the next chapter.

4. Check your answers against the correct answers located at the back of the book.

5. Follow all instructions outlined in each chapter. Take the time to read carefully. If you misunderstand any information you will continue to build on this misinformation.

Stan Aglow

CHAPTER 1

WHAT IS A BLUEPRINT?

The history of the blueprint and the evolution of blueprint reproduction must be known before delving into more complex issues.

BLUEPRINT REPRODUCTION

An architect or engineer spends many hours preparing a simple drawing. For this reason, the original drawing is very valuable, so instead of passing the drawing around from contractor to contractor, copies must be made. Reproducing each drawing by hand is not cost effective, so alternate methods have been devised.

Blueprinting. Originally, blueprint copies were made using a chemical method. This method consisted of a flat wooden surface surrounded by a wooden frame. Felt covered the wooden surface and a plate of glass attached to the wooden frame. Paper coated with a sunlight-sensitive chemical was then placed on the felt surface, coated side up. The original drawing was positioned over the paper,

right side up, and the plate of glass was secured firmly over both paper and original drawing. The entire assembly was placed in the sun for a period of time, then the chemically-treated paper was removed, washed in water, and hung out to dry. As the paper dried, it turned a dark blue. White lines, representing the original drawing, were reproduced onto this dark blue background with little or no noticeable distortion. This is how the name blueprint came into being, although this method is now obsolete. Figure 1-1 illustrates a blueprint of a duct section

Whiteprinting. A method now used quite often is called whiteprinting. This method consists of reproducing drawings with black lines onto white paper (while these prints are not blue, they are still called blueprints). Whiteprinting involves exposing

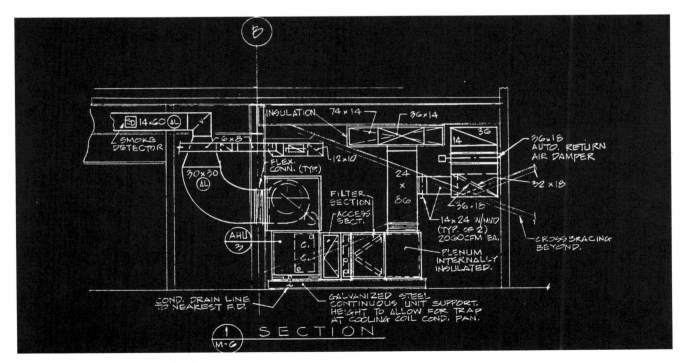

Figure 1-1. Blueprint of Duct Section

the original drawing and chemically-treated paper to a high-intensity fluorescent lamp. As in blueprinting, the original drawing is placed over the chemically-treated paper and then exposed to light. The ultraviolet rays from the lamp imprint the original drawing into the chemically-treated paper. At this point, the drawing cannot be seen on the chemically-treated paper. This paper must be exposed to ammonia vapors, which develop the original drawing on the paper. Figure 1-2 is the same duct section that was shown in Figure 1-1, except Figure 1-2 is a whiteprint, rather than a blueprint.

Photocopying. Photocopying is another method often used to reproduce blueprints. Photocopying is easy, as a simple copy machine is the only necessary tool. In this way, many copies of a drawing can be made, and the copy machine can reduce or enlarge a drawing to many sizes.

Photocopying is also advantageous when revisions need to be made to an original drawing. Instead of retracing a drawing, part of the photocopy can be cut out and redrawn on the photocopy, without ruining the original drawing.

Microfilming. Microfilming is used often, because drawings can be well preserved in this manner. Microfilm contains a tiny reproduction of the original drawing, so space is saved when storing the print. A special camera is needed for this process, in order to reduce the drawing to fit onto the microfilm. Once the film is mounted onto some sort of frame, the image can be reproduced, at full size, countless times.

Plotters. Still another method consists of using a copy machine, called a plotter, simultaneously with a computer system. This method involves the architect or engineer creating the original image on a computer (this process is called CAD, which stands for computer-aided design). This plotter copies the image from the computer and transfers the image to paper. The paper rolls forward as the plotter ink pens move horizontally across the paper. These ink pens come in a variety of colors. Figure 1-3 shows two types of plotters in use, and Figure 1-4 depicts a desktop plotter. Many of the drawings used throughout this book were created within a computer without the use of paper or pencil.

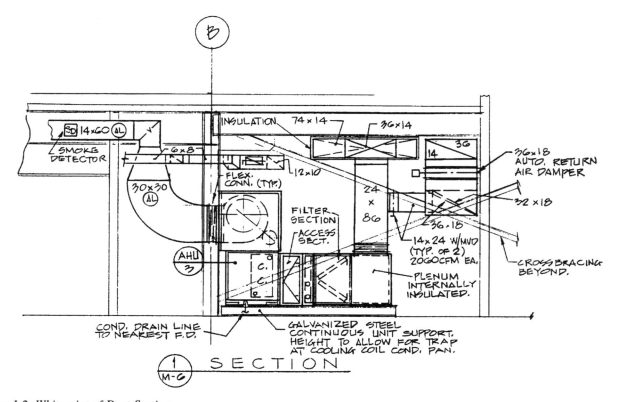

Figure 1-2. Whiteprint of Duct Section

Figure 1-4. Desktop Plotter. Courtesy, Houston Instrument.

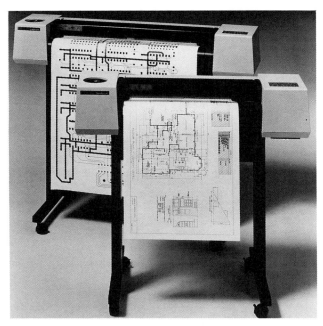

Figure 1-3. Plotters. Courtesy, Hewlett-Packard.

DAMAGED BLUEPRINTS

The original blueprint can sometimes become wrinkled, torn, or stained to the point where it can no longer be reproduced. The blueprint can always be redrawn, however, this is an expensive and time-consuming project. There are methods available that help renew the original blueprint to the point where it can be reproduced.

If the blueprint is stained, the stain can be "removed" using a filter placed over the lens of the camera in the photocopying machine. For the best results, it is necessary to match the color of the filter as closely as possible to the color of the stain; for example, a green stain would require a green filter over the camera. This method is often very successful.

Original blueprints can often be cleaned up by simply brushing a cotton ball soaked in a water-alcohol solution (using equal parts of water and alcohol) over the blueprint. The cotton ball should be squeezed to remove excess water-alcohol solution before applying it to the blueprint.

BLUEPRINTS IN CONSTRUCTION

The blueprint process starts when the owner of a proposed building hires an architect to design the building. The architect then prepares the design in blueprint format. Once the owner approves the blueprints, the architect selects several engineers to design the mechanical aspects of the building (i.e., structure, plumbing, heating, cooling). The engineers prepare more blueprints, detailing these mechanical aspects, and these blueprints, along with the architect's blueprints, make up a complete set of *working blueprints*.

Most cities require a building permit for new construction; therefore, the city must receive a set of the working blueprints. Once the city approves the blueprints, contractors bid for the opportunity to construct various aspects of the actual construction. Some projects require the bidding contractor to either leave a deposit or completely pay for the set of blueprints pertaining to that contractor's specialty (i.e., electrical, HVAC, etc.); the contractor never receives the entire set of working blueprints. The reason for the deposit is to eliminate contractors who may not be serious about performing the construction.

The contractors are then selected, and the blueprint becomes part of the legal document between the architect, engineers and contractors. The instructions and measurements contained in the blueprint must be followed in every detail, unless all parties involved approve a change. A misunderstood detail on the blueprint can lead to construction delays, additional costs or even a legal battle.

QUESTIONS FOR CHAPTER 1

Q-1   Why is the construction print called a blueprint?

A-1   _____

_____

_____

Q-2   What is one of the more common methods used to reproduce an original drawing?

A-2   _____

_____

_____

Q-3   Why is microfilm useful in blueprint reproduction?

A-3   _____

_____

_____

Q-4   How can damaged blueprints be restored?

A-4   _____

_____

_____

Q-5 What is the name of the complete set of blueprints?

A-5 _____

Q-6 Why is it necessary to understand what is drawn on the blueprint?

A-6 _____

_____

_____

Q-7 Will each subcontractor receive a complete set of construction prints?

A-7 _____

Q-8 What is a CAD program?

A-8 _____

_____

_____

Q-9 What is the title of the person who does the actual drawing of the blueprint?

A-9 _____

Q-10 How are CAD drawings transferred from the computer to paper?

A-10 _____

_____

_____

Q-11 Should changes be made to the blueprint by the contractor?

A-11 _____

Q-12 What procedure should be followed to make changes to the blueprint?

A-12 _____

_____

_____

CHAPTER 2

BLUEPRINT SYMBOLS AND WRITTEN LANGUAGE

All construction blueprints contain symbols and some form of written communication. The symbols represent a variety of items, from light switches to garage doors, and the written language can be in the form of abbreviations or actual construction terms. It is necessary to cover the symbols and language at this point, in order to understand the rest of the book.

SYMBOLS AND ABBREVIATIONS

Architects and engineers use symbols and abbreviations in blueprints in order to simplify their work. It would be extremely time-consuming to draw every component in a building as it appears to the eye, or to spell out every component. Also, if this were the case, the number of blueprints created would be tremendous.

Architects and engineers who are creating construction blueprints normally use symbols and abbreviations adopted by the United States of America Standards Institute. The symbols used are universal, ensuring all contractors using the blueprints will understand each symbol and abbreviation; however, many architects and engineers alter these symbols and abbreviations to suit their own tastes. Therefore, most drawings contain a legend or symbol list somewhere on the drawing, detailing the symbols or abbreviations used.

It is necessary to understand each symbol, which includes knowing the component or material it represents. While it may not be necessary to memorize each symbol, it is important to at least be able to identify which material or component the symbol represents.

Material Symbols. Material symbols represent the different materials that may be used when constructing a building. Figure 2-1 illustrates some of these material symbols and abbreviations.

Figure 2-2 shows how the material symbols are used in actual construction plans. This particular figure is an elevation view. In addition to the elevation view, the material symbols are also often represented on a floor plan, as in Figure 2-3. Material symbols can be combined and shown either on elevation views, or on floor plans.

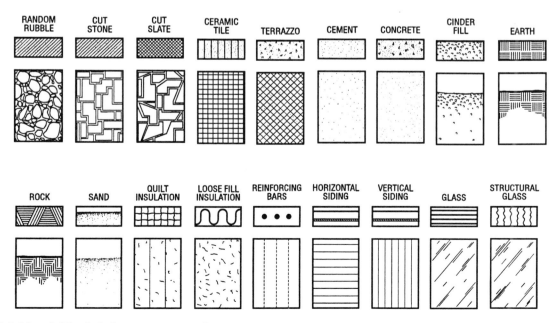

Figure 2-1. Material Symbols (cont. on next page)

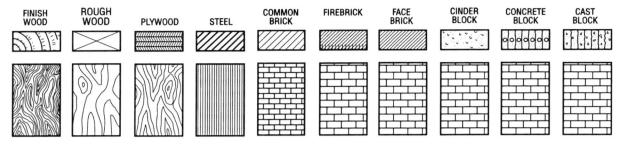

FINISH WOOD	ROUGH WOOD	PLYWOOD	STEEL	COMMON BRICK	FIREBRICK	FACE BRICK	CINDER BLOCK	CONCRETE BLOCK	CAST BLOCK

Figure 2-1. Material Symbols

Electrical Symbols and Schematics. Electrical drawings utilize symbols and abbreviations extensively. In an electrical drawing, it is necessary for the engineer to show, for example, the power outlets, wire sizes, and service equipment. Symbols are necessary to simplify the engineer's work. These symbols vary somewhat from schematics to other types of blueprints, and this will be obvious on the figures that follow.

The electrical symbols and abbreviations shown in Figure 2-4 are used on most electrical drawings. As the symbols are universal, the contractor has no problem understanding each symbol. Sometimes, the symbol is combined with an abbreviation, in order to further clarify a component. In electrical drawings, circles and squares usually denote an electrical load. The type of electrical load or electrical device is indicated by a letter placed inside.

Many of the symbols shown are similar to one another, and this can cause some confusion. Therefore, the easiest way to memorize the symbols is to first memorize a basic symbol, then look at the different lines or abbreviations that may be added to that basic symbol that cause the symbol to change its meaning.

When learning electrical symbols, it is helpful to look at an electrical supply company catalog. Then, it is possible to compare the actual component to the symbol. There are often similarities between the component and the symbol. The schematic is the engineer's blueprint of an electrical circuit. It differs from a builder's blueprint in that it does not show the actual location of the electrical devices used in the circuit, but only the electrical relationship of one device to another. Figure 2-5 shows some of the electrical schematic symbols.

HVAC Symbols. Heating, cooling, and air-conditioning units must be shown on HVAC drawings. Also, it is necessary to detail the ductwork and piping required for these systems. The piping symbols for HVAC systems is the same as for plumbing blueprints. Figure 2-6 shows the common HVAC symbols.

Plumbing Symbols. In addition to showing all the piping necessary for a particular building, plumbing blueprints must also include symbols indicating all fixtures, valves, and fittings. Figure 2-7 shows the various plumbing symbols used most often in plumbing blueprints.

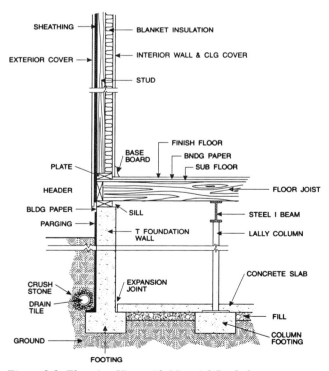

Figure 2-2. Elevation View with Material Symbols

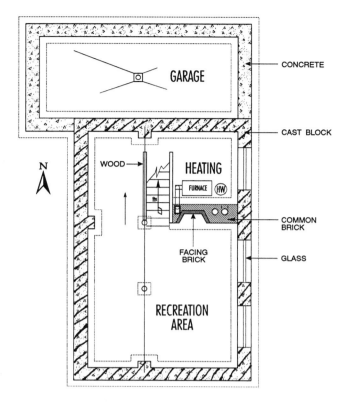

Figure 2-3. Floor Plan with Material Symbols

Elevation and Floor Plan Symbols. Elevation views and floor plans are the most common types of blueprints used. The symbols encountered between these two types can vary. In an elevation view, it is customary for the component to actually look like the component, rather than using a symbol. In a floor plan, this is not feasible, therefore, symbols are used more in this case. Figure 2-8 shows several common components in both the floor plan view and the elevation view.

Construction Symbols. Construction symbols illustrate the actual components required to support a structure. This helps the contractor to know exactly what kinds of beams, studs, tubes, etc., are necessary for a particular building. Figure 2-9 shows the different construction symbols used on blueprints.

Reference Symbols. Reference symbols are extremely important in identifying each blueprint. These symbols are found on virtually every blueprint and contain information concerning the direction in which the building is facing and sheet number of each blueprint. Figure 2-10 illustrates these reference symbols and Figure 2-11 shows how these reference symbols look in an actual blueprint.

WRITTEN LANGUAGE

As stated previously, it is necessary to understand the written language used on blueprints. Often, this language takes the form of abbreviations on the blueprints, and some of these abbreviations have already been illustrated. In addition to abbreviations, it is important for the contractor to understand the basic construction terms and synonyms for these terms. The synonyms are necessary to know, because in different parts of the country, certain materials are known by different names. Basically, written language on a blueprint is necessary to further describe the type and size of materials to be used, as well as the actual components involved.

Abbreviations. Capital letters are used for most abbreviations in blueprints. If an abbreviation can be mistaken for an actual word, then a period is placed after the abbreviation. For example, APT. is the abbreviation for apartment. Whether the construction term being abbreviated is singular or plural, the same abbreviation is used in either case. Figure 2-12 is a listing of some abbreviations.

Synonyms. Depending on where a person lives often determines what term that person will use when describing an object. Often, the architect, engineer, and contractor each use a different term to describe the same object. For example, what one person may call concrete, another person may call mortar. For this reason, it is necessary to know the common synonyms used in construction. Figure 2-13 lists some of the common synonyms used in construction.

NAME	ABBREV	SYMBOLS
FLUORESCENT LIGHT	LT FLUOR	
INCANDENSCENT LIGHT	LT	
WALL MOUNTED LIGHT	LT OUT WALL	
THERMOSTAT	T	T
EXIT LIGHT	X	
DUPLEX OUTLET	DUP OUT	
EMERGENCY LIGHT	E	E
FIRE ALARM	FA	
BELL	BL	
SOUND SYSTEM		
JUNCTION BOX	JUNC BX	J
SINGLE POLE SWITCH	S	S
DOUBLE POLE SWITCH	S_2	S_2
THREE WAY SWITCH	S_3	S_3
FOUR WAY SWITCH	S_4	S_4
CIRCUIT BREAKER	S_{CB}	S_{CB}
WATERPROOF SWITCH	S_{WP}	S_{WP}
AUTOMATIC DOOR SWITCH	S_D	S_D
PILOT LIGHT SWITCH	S_P	S_P
LOW-VOLTAGE SWITCH	S	S

NAME	ABBREV	SYMBOLS
LOW-VOLTAGE MASTER SWITCH	MS	MS
FUSED SWITCH	S_F	
MOTOR CONNECTION		
SINGLE OUTLET	S OUT	1
HEAVY-DUTY OUTLET 220 VOLTAGE	HVY DTY OUT	
REFRIGERATOR OUTLET	REF OUT	R
WATERHEATER OUTLET	WH OUT	WH
MOTOR OUTLET	M OUT	M
STRIP OUTLET	STP OUT	6"
GROUNDED OUTLET	GRD OUTLET	GR
FREEZER OUTLET	FR OUT	FR
BUZZER	BZR	
PUSH BUTTON	PB	
DIMMER SWITCH	DM SW	S_{DM}
FAN	F	
ELECTRIC HEATER	ELEC HTR	
POWER LINE SWITCH TO OUTLET	POW LN SW/OUT	S
SMOKE DETECTOR	SD	SD

Figure 2-4. Electrical Symbols and Abbreviations

NAME	SYMBOLS
PUSH BUTTON NO	
PUSH BUTTON NC	
LOW PRESSURE SWITCH	
HIGH PRESSURE SWITCH	
COOLING THERMOSTAT	
HEATING TEMPERATURE	
CONTACTS NO	
CONTACTS NC	
FLOW SWITCH	
FLOW SWITCH	

NAME	SYMBOLS
FLOAT SWITCH	
FLOAT SWITCH	
CAPACITOR	
RESISTOR	
OVERLOAD	
FUSE	
RELAY	
THERMOCOUPLE	
CROSSOVER	

NAME	SYMBOLS
CONNECTION	
DISCONNECT	
FUSED DISCONNECT	
GROUND	
TIME DELAY	
SOLENOID	
LOAD	
MOTOR	

Figure 2-5. Electrical Schematic Symbols

NAME	SYMBOLS
CEILING DIFFUSER	
SUPPLY AIR DUCT	
RETURN AIR DUCT	
TURNING VANE IN DUCT	
LINEAR DIFFUSER	
RETURN CEILING DIFFUSER	
RECTANGULAR DUCT 1st Figure width 2nd Figure depth	24 x 8
EXHAUST FAN	
FAN COIL	

NAME	SYMBOLS
INSULATED DUCT	
DUCT SIZE CHANGE	
CEILING-DUCT OUTLET	
WARM-AIR SUPPLY	WA
DUCT SIZE & FLOW DIRECTION	
DUCT RISING	R R
COMPRESSOR	
RECEIVER	
FUEL-OIL TANK	OIL

NAME	SYMBOLS
HEAT REGISTER	R
ROOM AIR CONDITIONER	RAC
COLD-AIR RETURN	CA
HEAT OUTLET	
RADIATOR	RAD
HEAT PUMP	HP
HUMIDISTAT	H
GAS OUTLET	g

Figure 2-6. HVAC Symbols

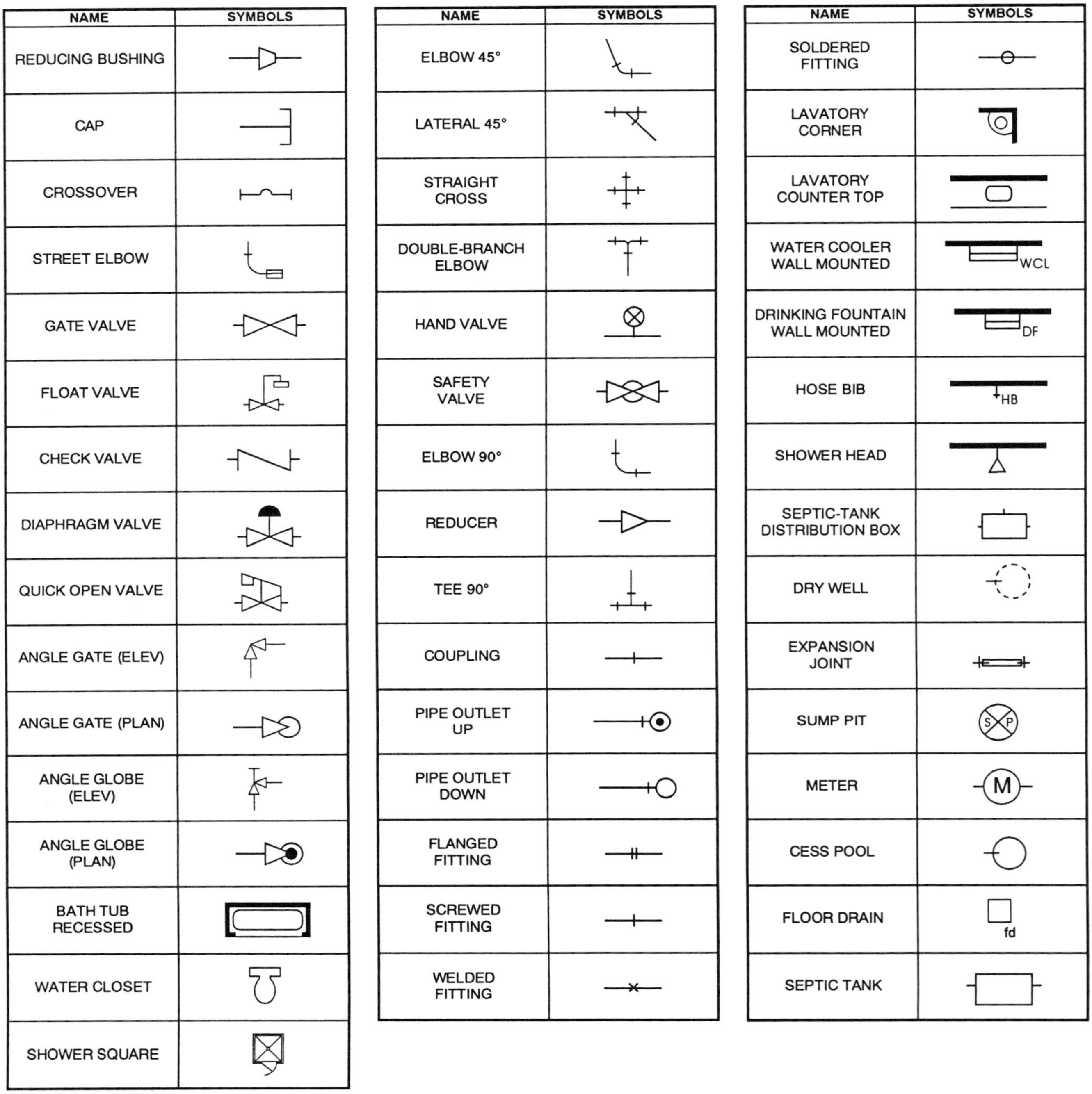

NAME	SYMBOLS
REDUCING BUSHING	
CAP	
CROSSOVER	
STREET ELBOW	
GATE VALVE	
FLOAT VALVE	
CHECK VALVE	
DIAPHRAGM VALVE	
QUICK OPEN VALVE	
ANGLE GATE (ELEV)	
ANGLE GATE (PLAN)	
ANGLE GLOBE (ELEV)	
ANGLE GLOBE (PLAN)	
BATH TUB RECESSED	
WATER CLOSET	
SHOWER SQUARE	

NAME	SYMBOLS
ELBOW 45°	
LATERAL 45°	
STRAIGHT CROSS	
DOUBLE-BRANCH ELBOW	
HAND VALVE	
SAFETY VALVE	
ELBOW 90°	
REDUCER	
TEE 90°	
COUPLING	
PIPE OUTLET UP	
PIPE OUTLET DOWN	
FLANGED FITTING	
SCREWED FITTING	
WELDED FITTING	

NAME	SYMBOLS
SOLDERED FITTING	
LAVATORY CORNER	
LAVATORY COUNTER TOP	
WATER COOLER WALL MOUNTED	WCL
DRINKING FOUNTAIN WALL MOUNTED	DF
HOSE BIB	HB
SHOWER HEAD	
SEPTIC-TANK DISTRIBUTION BOX	
DRY WELL	
EXPANSION JOINT	
SUMP PIT	S P
METER	M
CESS POOL	
FLOOR DRAIN	fd
SEPTIC TANK	

Figure 2-7. Plumbing Symbols

NAME	PLAN VIEW	ELEVATION
WINDOW	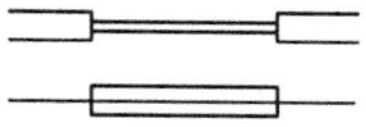	
DOORS	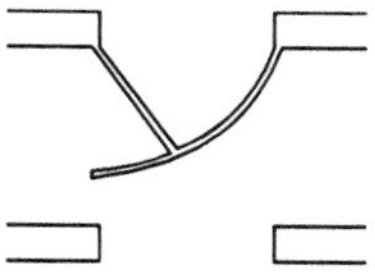	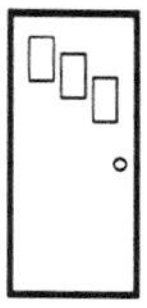
SIDING		
THERMOSTAT		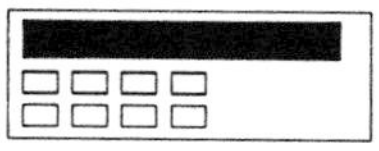
OUTLET	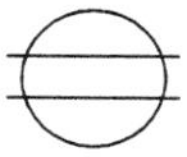	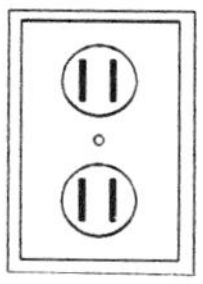
SWITCH	 S	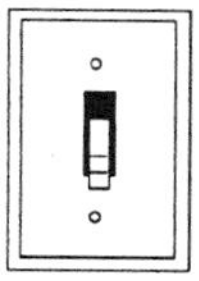
BREAKER BOX		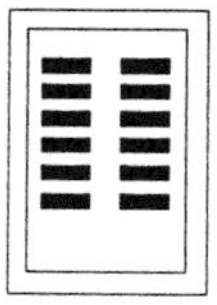

Figure 2-8. Elevation and Floor Plan Symbols

NAME	SYMBOLS
ROUGH WOOD	
FINISH WOOD	
PLYWOOD	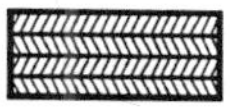
METAL CURTAIN WALL	
METAL STUD WALL	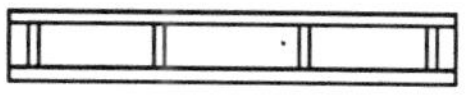
STRUCTURAL TEE BEAM	 T
ANCHOR BOLT	
PLATE	 PL
COLUMN	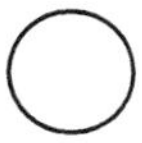
BULB TEE BEAM	 BT
LALLY COLUMN	
STEEL TUBE	
SQUARE BAR	
ANGLE BEAM	
I BEAM	
WIDE FLANGE BEAM	 WF
CHANNEL BEAM	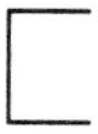

Figure 2-9. Construction Symbols

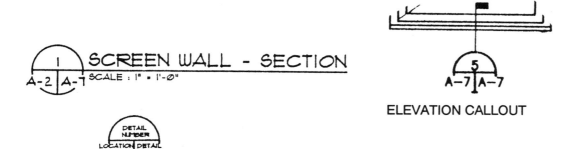

SCREEN WALL - SECTION
SCALE : 1" = 1'-0"
A-2 | A-7

ELEVATION CALLOUT
A-7 | A-7
5

DETAIL NUMBER
LOCATION DETAIL
SHEET SHEET
DETAIL SYMBOL

SECTION DETAIL CALLOUT

NORTH

PLOT DIRECTION

Figure 2-10. Reference Symbols

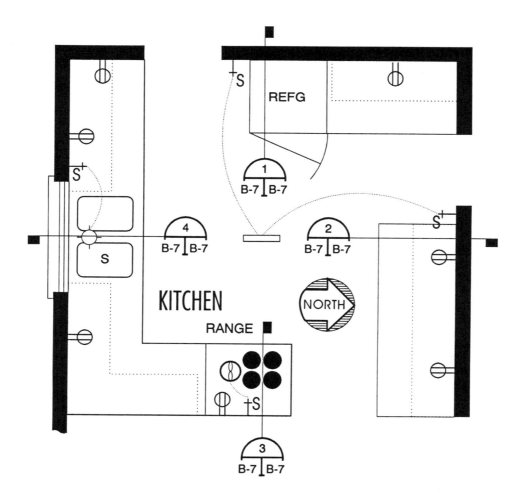

REFG

KITCHEN

RANGE

NORTH

1
B-7 | B-7

4
B-7 | B-7

2
B-7 | B-7

3
B-7 | B-7

Figure 2-11. Reference Symbols in a Blueprint

Air conditioning: A/C	**I beam:** I
Bathroom: B	**Insulate:** INS
Blueprint: BP	**Kilowatt:** KW
British thermal units: Btu	**Kilowatt hour:** KWH
Circuit breaker: CIR BKR	**Laundry:** LAU
Closet: CL	**Opening:** OPNG
Entrance: ENT	**Plumbing:** PLMB
Floor: FL	**Radiator:** RAD
Floor drain: FD	**Refrigerator:** REF
Foot: (ʹ) FT	**Register:** REG
Garage: GAR	**Valve:** V
Hall: H	**Washing machine:** WM
Hot water: HW	**Water heater:** WH

Figure 2-12. Common Abbreviations

Aeration: support	**Flush valve:** float valve
Auxiliary window: storm window	**Ground line:** grade
Baseboard: finish board	**Heat transmission:** conductor
Basement: cellar	**Laundry:** utility room
Bibs: faucets	**Rain drainage:** drainage pipe
Caulking: sealer	**Sealer:** caulking
Column: post	**Skeleton:** framing
Damper: flue control	**Smokestack:** chimney, flue
Dehydration: evaporation	**Taps:** faucets
Dry well: reservoir	**Thermostat:** automatic temperature control
Fan: blower	**Tread:** step
Flashing: vapor barrier	**Tubes:** ducts
Flue: chimney	**Wall:** partition

Figure 2-13. Synonyms

QUESTIONS FOR CHAPTER 2

Q-1   Why do architects and engineers use symbols in their work?

A-1   _____

_____

_____

Q-2   Draw the common material symbols for each of the following:

A-2   earth -

rock-

Q-3   Draw the common electrical symbols for each of the following:

A-3   thermostat-

motor connection-

Q-4   What is the abbreviation for each of the following:

A-4   smoke detector-

exit light-

Q-5 Draw the symbols for each of the following:

A-5 exhaust fan-

heat register-

Q-6 Draw the symbols for each of the following:

A-6 supply air duct -

heat pump -

Q-7 In a plumbing blueprint, what do the symbols normally represent?

A-7 _____

_____

_____

Q-8 Draw the symbols for each of the following:

A-8 coupling -

expansion joint-

Q-9 Which types of blueprints use symbols most often?

A-9 _____

_____

_____

Q-10 Draw the symbol for a steel tube.

A-10

Q-11 Why is it necessary to know construction synonyms?

A-11 _____

_____

_____

Q-12 What is the common abbreviation for a British Thermal Unit?

A-12 _____

Q-13 Name a synonym for the following:

Q-13 faucet-

wall-

CHAPTER 3
THE USE OF LINES IN A BLUEPRINT

In construction blueprints there are many types of lines in use, including center lines, phantom lines, electrical lines, etc.

BLUEPRINT LINES

A line in a blueprint is considered a symbol. These lines are similar to the symbols discussed in the previous chapter, in that lines come in many styles and weights. The style and weight of the line depends on the particular construction method or meaning employed in a certain blueprint.

Lines can be tricky symbols, because their meanings change from field to field, or even engineer to engineer. For example, a broken line in a mechanical drawing (- - - - -) can represent an object that is located behind another object on the blueprint, therefore it is invisible. In an electrical blueprint, the broken line can represent wiring that is installed in the field by the service technician or electrician. A thin broken line can indicate low voltage wiring in an electrical blueprint, and a thick broken line can indicate high voltage. For this reason, it is necessary to include a legend in every blueprint, because the slightest difference in thickness or style can entirely change the meaning of the line.

Figure 3-1 shows the more common lines used on blueprints. As stated previously, however, line definitions do change from field to field. Figure 3-2 illustrates how lines are used in actual blueprints.

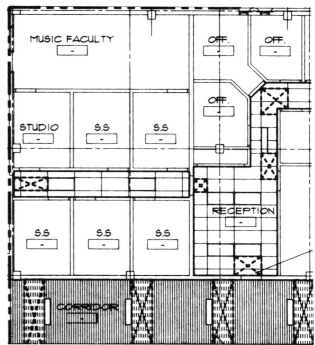

Figure 3-2. Lines in Actual Blueprints. Courtesy, Northwest Blueprint and Supply Company.

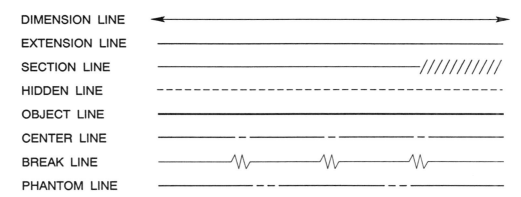

Figure 3-1. Blueprint Lines

In order to fully understand Figure 3-1, a definition of each line is necessary.

A *dimension line* is a thin, unbroken line, with an arrow on each end of the line. A measurement is usually placed upon the line in inches, feet, degrees, meters, etc. This type of line shows the measurements of a building from one point to another.

An *extension line* extends from the edge of the building, and this is the line to which the dimension arrows point.

A *section line* separates different materials in a sectional drawing. Another symbol then differentiates each building material in the blueprint. It is easiest to think of a section line as a flat cutting plane passing through a solid. The cut edges will be shaded with a cross hatch to show thickness.

A *hidden line* shows objects or areas that are hidden by other objects or areas. Floor plans often use hidden lines to show objects such as wall cabinets, beams, and arches, as these objects appear above the section line.

An *object line* defines the boundaries of the building. This line includes the exterior and interior walls, patios, decks and driveways. This type of line is the most easily visible on a blueprint.

A *center line* identifies the center of symmetrical objects, including beams, floors, wall studs, windows, doors, etc. The exact location of a building component can be determined by measuring from the center line. The center line can also be identified by the letters (CL) or (C).

A *break line* indicates the entire area is not represented on the blueprint. This type of line is used when a particular component or feature continues for a long length. For example, piping that remains the same for a considerable distance often uses break lines. A dimension line is often used in conjunction with the break line in order to show the total length of the object. There are two types of break lines: (1) a long, straight line broken by short sections of wavy lines, used for long, straight breaks and (2) a wavy line used for small breaks.

A *phantom line* indicates an object that is not really there, or else an object that must be removed in order for new construction to take place. For example, construction additions or optional locations for fixtures or moveable walls may be indicated with a phantom line.

Stair Line. Buildings that consist of two or more floors usually have elevators or stairs, in order for the occupants to move throughout the building. For buildings under three floors, stairs are normally used. Stairs in blueprints are represented with lines indicating the direction of the stairs (whether the stairs go up or down). For example, stairs leading from the first floor to a basement are labeled down, which indicates the stairs lead to the basement. Stairs leading from a first-floor to a second floor are labeled up, which indicates the stairs lead to an upper floor. Figure 3-3 shows examples of stair lines.

A break line is often used in conjunction with the stair line. This indicates the stairs continue, even though the blueprint stops.

Stairs in a building may also include a landing, when the length of the stairs is considerable. When a landing is used, the stairs take on an L-shape, illustrated in Figure 3-4. The stair lines then indicate whether the stairs go up or down.

Other Lines. The previous chapter illustrated some of the lines used in other fields; for example, mechanical and electrical plans use the lines to represent piping or wiring. Figures 3-5 and 3-6 show some the lines used in electrical wiring and plumbing blueprints.

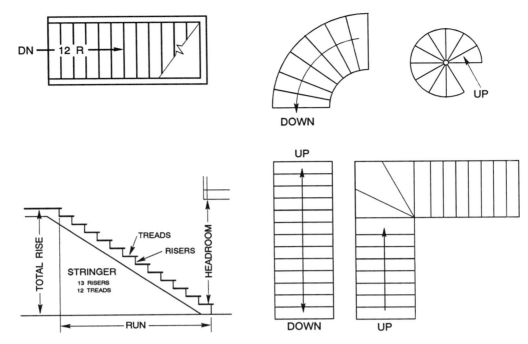

Figure 3-3. Stair Lines

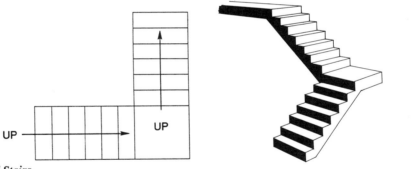

Figure 3-4. L-Shaped Stairs

CONCEALED CONDUIT CEILING AND WALLS	————————————————
CONCEALED CONDUIT FLOORS AND WALLS	– – – – – – – – – – – – – –
INDICATES THE NUMBER OF CONDUCTORS	—— ||||||| ————————
METALLIC ARMORED CABLE, FLEXIBLE	——————— X ———————
TELEPHONE CONDUIT	——————— T ———————
TV ANTENNA CONDUIT	——————— TV ———————
SOUND SYSTEM	——————— S ———————

Figure 3-5. Lines for Electrical Wiring

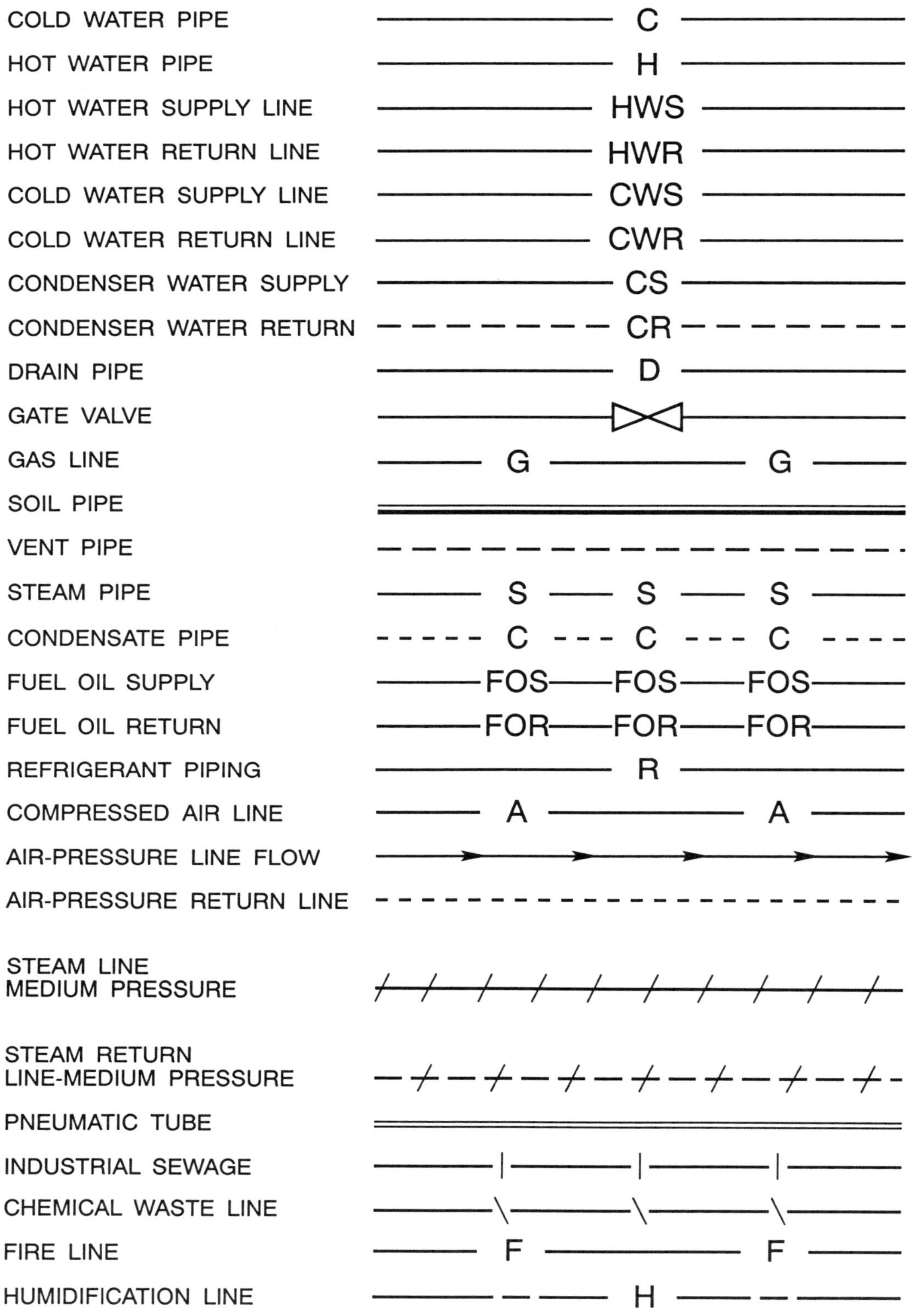

COLD WATER PIPE	——————— C ———————
HOT WATER PIPE	——————— H ———————
HOT WATER SUPPLY LINE	——————— HWS ———————
HOT WATER RETURN LINE	——————— HWR ———————
COLD WATER SUPPLY LINE	——————— CWS ———————
COLD WATER RETURN LINE	——————— CWR ———————
CONDENSER WATER SUPPLY	——————— CS ———————
CONDENSER WATER RETURN	– – – – – – CR – – – – – –
DRAIN PIPE	——————— D ———————
GATE VALVE	———————▷◁———————
GAS LINE	——— G ——————— G ———
SOIL PIPE	═══════════════════
VENT PIPE	– – – – – – – – – – – – – – – ·
STEAM PIPE	——— S ——— S ——— S ———
CONDENSATE PIPE	– – – – C – – – C – – – C – – – –
FUEL OIL SUPPLY	———FOS——FOS——FOS———
FUEL OIL RETURN	———FOR——FOR——FOR———
REFRIGERANT PIPING	——————— R ———————
COMPRESSED AIR LINE	——— A ——————— A ———
AIR-PRESSURE LINE FLOW	——→———→———→———→
AIR-PRESSURE RETURN LINE	– – – – – – – – – – – – – – – – –
STEAM LINE MEDIUM PRESSURE	⁄—⁄—⁄—⁄—⁄—⁄—⁄—⁄—⁄
STEAM RETURN LINE-MEDIUM PRESSURE	—⁄—⁄—⁄—⁄—⁄—⁄—⁄—
PNEUMATIC TUBE	═══════════════════
INDUSTRIAL SEWAGE	———|———|———|———
CHEMICAL WASTE LINE	———\———\———\———
FIRE LINE	——— F ——————— F ———
HUMIDIFICATION LINE	———— –– —— H —— –– ————

Figure 3-6. Lines in Plumbing

QUESTIONS FOR CHAPTER 3

Q-1 What is a section line?

A-1 _____

_____

_____

Q-2 What is an object line?

A-2 _____

_____

_____

Q-3 Draw a dimension line.

A-3

Q-4 What is the importance of a center line?

A-4 _____

_____

_____

Q-5 Draw a center line.

A-5

Q-6   When might a break line be used?

A-6   _____

_____

_____

Q-7   Why should the line legend always be included?

A-7   _____

_____

_____

Q-8   What might a phantom line represent?

A-8   _____

_____

_____

Q-9   Why are stair lines important?

A-9   _____

_____

_____

Q-10 Why are lines necessary in blueprints?

A-10 _____

_____

_____

CHAPTER 4
SCALES

It is impossible for a building to be drawn to full size; therefore, it is necessary for the architect or engineer to create a scaled-down version of the building on paper; this is called drawing to scale.

SCALES

There are three basic types of scales that can be used for construction blueprints: the architect's scale, the engineer's scale, and the metric scale. All three scales are used to convey information concerning the dimensions of the building to those constructing the building.

Architect's Scale. When drawing a blueprint to scale, the architect reduces all the lines in the drawing by the same ratio. For example, an architect may use a 1/4-inch line in the drawing to represent an actual 1-foot segment of the building. Dimensions on the drawing should be thought of in terms of the actual dimensions of the building. The scale is always shown on the blueprint; for example, 1/4 inch = 1 foot will appear somewhere on the blueprint if that is the scale being used.

Most architects use a certain distance (i.e., 1/4 inch, 1/2 inch, etc.) to represent 1 foot. The distance the architect selects is then divided into 12 equal sections, each of which then equals 1 inch. For example, an

architect may use a 1-inch measurement on a blueprint to represent 1 actual foot in the building. The 1-inch measurement is then subdivided into 12 equal parts, so each 1/12 inch on the blueprint would represent 1 actual inch in the building.

In order to create a reduced drawing, the architect uses a scale or ruler as shown in Figure 4-1. This scale may either look like a regular ruler, or it may be in the shape of a triangle. On the triangular scale, there are two scales available on each surface of the triangle, one scale reads left to right, and the other scale reads right to left. For example, the 1/4-inch = 1-foot scale may start at one end, and the 1/8-inch = 1-foot scale may start from the other. Figure 4-2 shows these two different scales on one surface of the scale.

All blueprints do not have to be drawn to scale. A drawing marked *not to scale* is normally used to show some unusual feature or installation procedure. This type of drawing may be drawn to a different scale than that used in the rest of the blueprints, and the

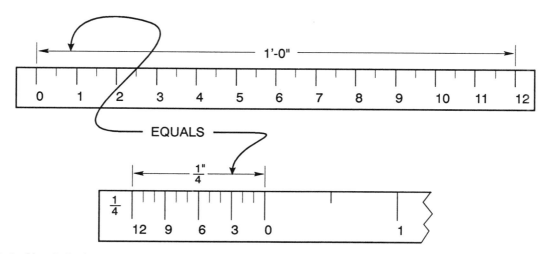

Figure 4-1. Architect's Scale

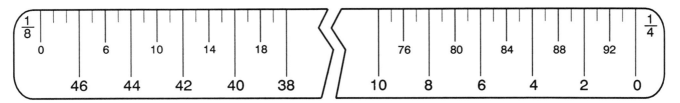

Figure 4-2. Combination Scales

dimension measurements are given as full size. Figure 4-3 is an example of a not-to-scale drawing.

When architects create drawings, they must make sure that the dimensions given in the blueprint are clear, so the contractor can plainly see the correct dimensions of each building component. In order for these dimensions to be clear, the architects must draw the dimension lines away from the building lines. The architects must also make sure that all blueprints (i.e., sectional, elevation, etc.) match with one another. Finally, it is important for the architects to show wall thickness, and when masonry openings are used, they should be labeled M.O., indicating an exact opening is necessary. Figure 4-4 is an example of a typical drawing created to an architect's scale.

Engineer's Scale. When creating site plans, an engineer's scale (also called a civil engineer's scale) is most often used. This scale differs from the architect's scale in that each inch in an engineer's scale is divided into 10, 20, 30, 40, 50, or 60 equal units. These units can be inches, feet, yards, miles, or whatever else the engineer desires. The unit chosen usually depends on the drawing size; however, in all cases, the unit chosen must be divisible by a power of 10. For example, the engineer's scale may be 1 inch = 100 feet, 1 inch = 1000 feet, etc. Figure 4-5 shows various engineers' scales.

When engineers create site plans, they place the engineer's scale along the direction of the measurement needed, for example, the property line.

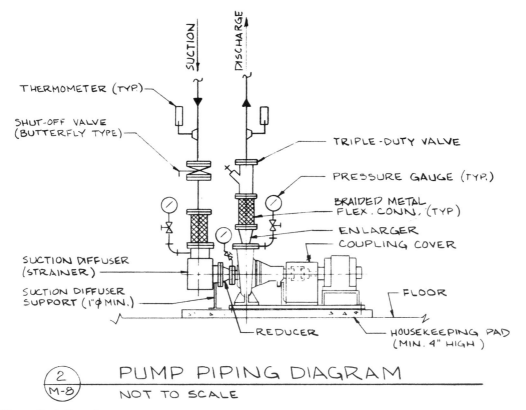

Figure 4-3. Not-to-Scale Drawing

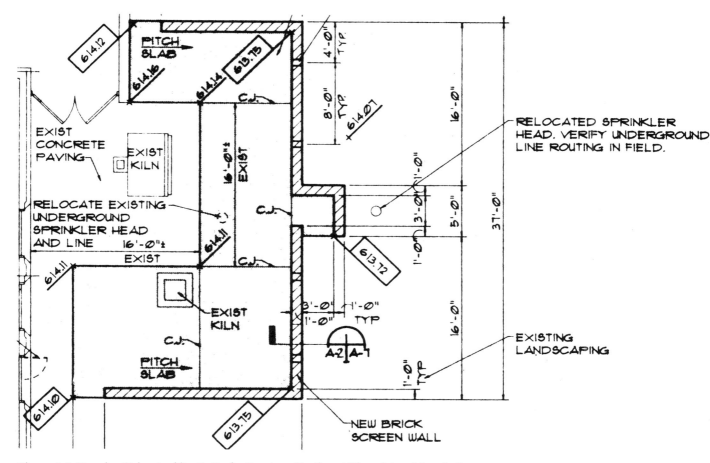

Figure 4-4. Drawing Using Architect's Scale. Courtesy, Northwest Blueprint and Supply Company.

If the drawing is already created, it is possible to obtain the length of the property line by simply reading the measurement on the scale. If a drawing needs to be created, the scale can be used to transfer the dimensions of the actual property to the drawing. The actual property dimensions (or whatever else is being measured) must always be shown on the drawing.

As with the architect's scale, the object being drawn should be thought of as full size, using the engineer's scale. For example, if the scale is 1 inch = 20 feet and the measurement on the drawing is 3 inches, the measurement is considered to be 60 feet (3 x 20 = 60).

There is another type of engineer's scale called the mechanical engineer's scale. On this type of scale, there is a main unit on the end which represents 1 inch. This inch is then divided into fractions of an inch, and these fractions are typically in multiples of 1/2; for example, 1/2, 1/4, 1/8, etc. Odd fractions, such as 1/3, 1/5, etc., are rarely used. Figure 4-6 is an example of a mechanical engineer's scale.

Metric Scale. Architectural drawings often use a metric scale. The metric scale uses a meter (m) to measure distance, the kilogram (kg) to measure weight, and the liter (L) to measure volume. Prefixes representing multiple meters are often attached; for example, a dekameter equals 10 meters, a hectometer equals 100 meters, and a kilometer equals 1000 meters. There are also prefixes that represent subdivisions of meters, including a decimeter, which equals 1/10 of a meter, a centimeter, which equals 1/100 of a meter, and a millimeter, which equals 1/1000 of a meter. The millimeter is the most commonly used subdivision. Figure 4-7 shows meter subdivisions.

When floor plans use the metric system, the measurements are usually carried out three decimal points when meters are used. Figure 4-8 illustrates a floor plan using the metric scale. The measurement does not have to be carried out three decimal points in a sectional drawing, as shown in Figure 4-9.

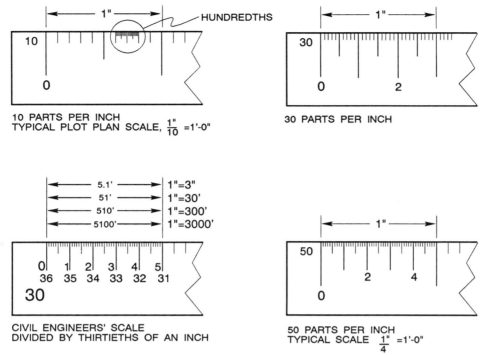

10 PARTS PER INCH
TYPICAL PLOT PLAN SCALE, $\frac{1"}{10}$ =1'-0"

30 PARTS PER INCH

CIVIL ENGINEERS' SCALE
DIVIDED BY THIRTIETHS OF AN INCH

50 PARTS PER INCH
TYPICAL SCALE $\frac{1"}{4}$ =1'-0"

Figure 4-5. Various Engineers' Scales

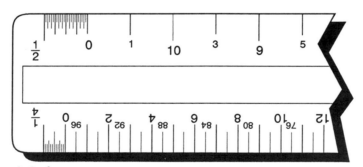

Figure 4-6. Mechanical Engineer's Scale

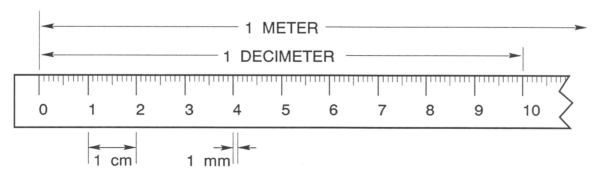

Figure 4-7. Meter Subdivisions

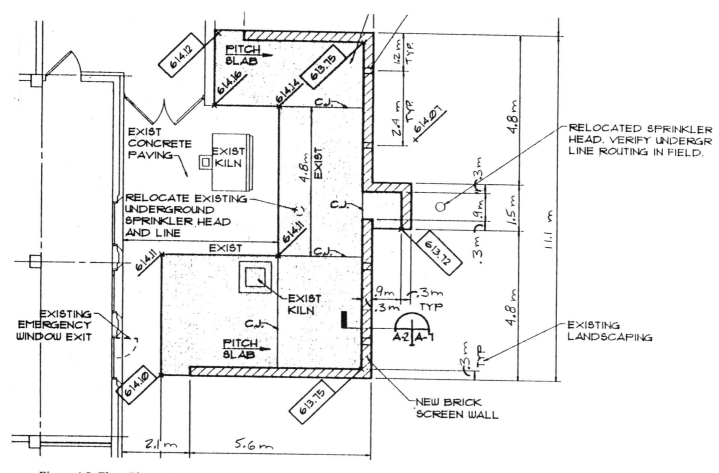

Figure 4-8. Floor Plan Using Metric Scale. Courtesy, Northwest Blueprint and Supply Company.

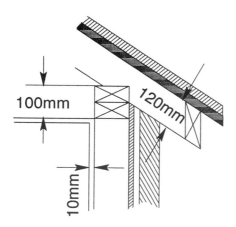

Figure 4-9. Sectional Drawing Using Metric Scale

The main difference between a metric scale and an architect's scale is that the metric scale is based on increments of 10, rather than increments of 12, as used in an architect's scale. As in the other scales, the unit of measurement chosen depends on the size of the drawing as compared to the actual size of the object being drawn.

Even though the metric scale is used in the United States, some components are not available in metric sizes. Therefore, it is often necessary to convert the dimension and distance measurements used in the U.S. to the metric scale. Figure 4-10 shows a standard metric conversion table.

Architects sometimes use dual dimensions in a drawing. This simply means that the metric scale and the standard U.S. dimensions are both shown. This eliminates the need to convert the dimensions. If dual dimensions are not used, a metric conversion chart often accompanies the drawing. Figure 4-11 illustrates dual dimensions in a drawing.

	IF YOU KNOW:	MULTIPLY BY:	TO FIND:
MASS	OUNCES	28.0	GRAMS
	POUNDS	0.45	KILOGRAMS
	SHORT TONS	0.9	MEGAGRAMS (METRIC TONS)
	GRAMS	0.035	OUNCES
	KILOGRAMS	2.2	POUNDS
	MEGAGRAMS (METRIC TONS)	1.1	SHORT TONS
TEMPERATURE	DEGREES FAHRENHEIT	5/9 (AFTER SUBTRACTING 32)	DEGREES CELSIUS
	DEGREES CELSIUS	9/5 (THEN ADD 32)	DEGREES FAHRENHEIT
LENGTH	INCHES	25.0	MILLIMETERS
	FEET	30.0	CENTIMETERS
	YARDS	0.9	METERS
	MILES	1.6	KILOMETERS
	MILLIMETERS	0.04	INCHES
	CENTIMETERS	0.4	INCHES
	METERS	1.1	YARDS
	KILOMETERS	0.6	MILES
AREA	SQUARE INCHES	6.5	SQUARE CENTIMETERS
	SQUARE FEET	0.09	SQUARE METERS
	SQUARE YARDS	0.8	SQUARE METERS
	SQUARE MILES	2.6	SQUARE KILOMETERS
	ACRES	0.4	SQUARE HECTOMETERS (HECTARES)
	SQUARE CENTIMETERS	0.16	SQUARE INCHES
	SQUARE METERS	1.2	SQUARE YARDS
	SQUARE KILOMETERS	0.4	SQUARE MILES
LIQUID VOLUME	OUNCES	30.0	MILLILITERS
	PINTS	0.47	LITERS
	QUARTS	0.95	LITERS
	GALLONS	3.8	LITERS
	MILLILITERS	0.034	OUNCES
	LITERS	2.1	PINTS
	LITERS	1.06	QUARTS
	LITERS	0.26	GALLONS

Figure 4-10. Standard Metric Conversion Table

SCALE SIMILARITIES

All the scales discussed in this chapter are distinctly different. However, all scales can either be open divided or fully divided.

An open-divided scale has the main units numbered along the entire length of the scale. This type of scale then has a fully-subdivided extra unit that is located at one end of the scale. This section shows the fractional properties of the main unit. Often, open-divided scales have two measuring systems on one side, as discussed in the architect's scale. Figure 4-12 illustrates an open-divided scale.

When reading an open-divided scale, it is important for the user to always start at the zero line and not at the beginning of the fully subdivided section. If additional inches or feet are then needed, these can be added from the subdivided section. Figure 4-13 shows an open-divided scale being used to find a specific measurement.

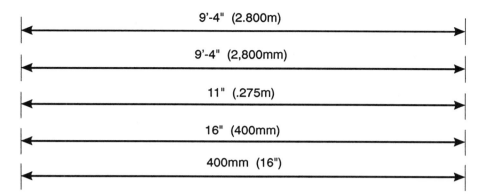

Figure 4-11. Dual Dimensions

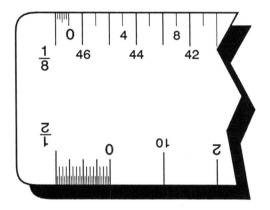

Figure 4-12. Open-Divided Scale

A fully-divided scale fully subdivides each main unit along the entire length of the scale. This is beneficial in that several values can be calculated at one time, instead of jumping back and forth between a fully-divided section and the rest of the scale. Figure 4-14 is an example of a fully-divided scale.

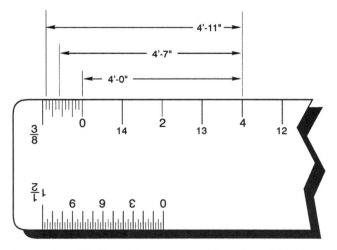

Figure 4-13. Finding a Measurement with an Open-Divided Scale

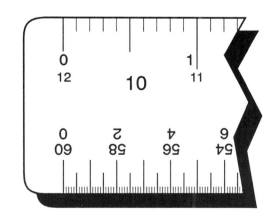

Figure 4-14. Fully-Divided Scale

QUESTIONS FOR CHAPTER 4

Q-1   Why are scales necessary in construction drawings?

A-1   _____

_____

_____

Q-2   Describe an architect's scale.

A-2   _____

_____

_____

Q-3   On a triangular scale, what is the maximum number of different scales available?

A-3   _____

Q-4   When are not-to-scale drawings used?

A-4   _____

_____

_____

Q-5 If a building is to be 50 feet long and is drawn to a scale of 1/8 inch = 1 foot, how long will the building actually measure on the drawing?

A-5 _____

Q-6 What guidelines do architects use to ensure their drawings are clear?

A-6 _____

_____

_____

Q-7 How does an engineer's scale differ from an architect's scale?

A-7 _____

_____

_____

Q-8 A site plan contains a legend that reads 1/4 inch = 30 feet. The measurement of a certain line on the site plan is 2 inches. What is the actual measurement?

A-8 _____

Q-9 What do the following measurements measure?

A-9 meter -

kilogram -

liter -

Q-10 What is the difference between a metric scale and an architect's scale?

A-10 _____

_____

_____

Q-11 What are dual dimensions?

A-11 _____

_____

_____

Q-12 What is an open-divided scale?

A-12 _____

_____

_____

Q-13 What is a fully-divided scale?

A-13 _____

_____

_____

CHAPTER 5

BLUEPRINT STYLES

Architectural structures can be depicted using several different blueprint styles. These styles are then applied to the different blueprint categories, which are discussed in the next chapter.

BLUEPRINT STYLES

There are several different types of blueprint styles: pictorial, orthographic, and diagrams or schematics. Each of these styles is used to convey information and further simplify the design of the building to the contractors who must construct the building.

Pictorial. In this type of drawing, the object is drawn three-dimensionally on a regular piece of drawing paper, so several faces of the object are shown at one time. Pictorial drawings are made for those people who may not be trained to read blueprints, or else they may be used in conjunction with orthographic drawings. There are several pictorial variations, and these include *perspective, isometric* and *oblique*.

The perspective drawing is a realistic drawing, in that it is drawn to represent the actual configuration of particular objects, as they appear to the eye. Figure 5-1 is an example of a perspective drawing using cabinets. Perspective drawings are not technically accurate, and they are often used to detail certain features of a particular design. The buildings can be distorted to highlight these certain features, which is why they are not technically accurate. The perspective drawing is often used in sales, especially when the potential buyer has a choice of two or more types of house exterior, even though the floor plan may be the same in all the houses.

In the perspective drawing, the horizontal front and rear lines are drawn parallel to each other, and the vertical front and rear lines are drawn parallel to each other. Depth is achieved by drawing diagonal lines toward an imaginary point, called the vanishing point, on the horizon. For example, when looking down

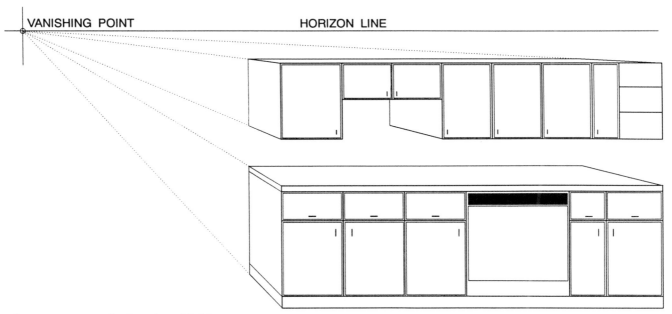

Figure 5-1. Perspective Drawing of Cabinets

train tracks, there is a point in the distance where the tracks look like they join together and then vanish. The perspective drawing tries to capture this type of view on a flat sheet of paper.

Figure 5-2 is a perspective drawing of a house. In this drawing, the roof line and the foundation line are extended to the same point on the horizon, thus the drawing is in perspective. Where these lines meet on the horizon is the vanishing point. The vanishing point is always located on the horizon line. The architect creates a perspective drawing based on this horizon line. For example, when the building is placed below the horizon line, the building looks like it is below the line of sight and when the building is placed above the horizon line, the structure looks as if it is at eye level. Figure 5-3 illustrates the relationship of the horizon line to how the building is drawn.

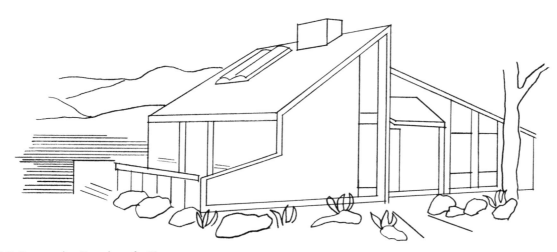

Figure 5-2. Perspective Drawing of a House

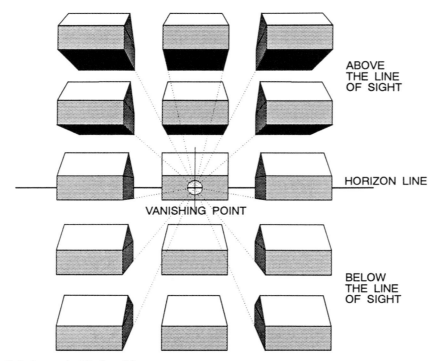

Figure 5-3. Building in Relation to the Horizon Line

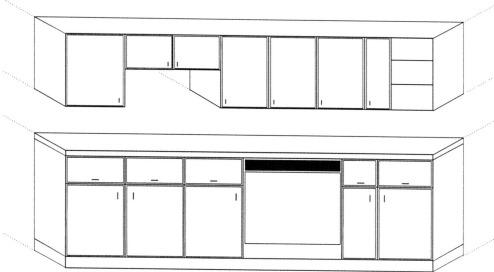

Figure 5-4. Isometric Drawing of Cabinets

The perspective drawing most closely resembles an actual photograph of the exterior or interior of the building. Unlike the isometric or oblique drawings, which appear distorted, the diagonal lines are not parallel to each other and appear not to be distorted.

The isometric style shows two sides and a top or bottom of the building or object. Figure 5-4 is an example of an isometric drawing. All the lines in an isometric drawing are foreshortened. Figure 5-5 is an isometric drawing of a cube. In this drawing, the edges are known as isometric axes, and the three surfaces showing are called isometric planes. The lines that are parallel to the isometric axes are called isometric lines.

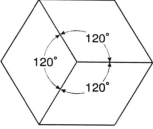

Figure 5-5. Isometric Drawing of a Cube

The isometric style of pictorial drawings is preferred over the perspective and oblique styles, because the isometric drawing may be drawn to scale, just like floor plans. For example, Figure 5-6 is an example of a plumbing blueprint drawn in the isometric style.

An oblique drawing is much the same as the isometric drawing, in that one side of the object is drawn to represent its exact appearance. The other exposed sides are represented by parallel lines that are drawn at the same angle, usually 30 to 45 degrees. These depth lines give the drawing a distorted view. In an oblique drawing, the lines are shortened to give a somewhat exact representation of the object being drawn. Therefore, the drawing is not drawn to scale. Figure 5-7 is an example of cabinets drawn in the oblique style.

It is important to understand the differences between the perspective, isometric, and oblique drawings. Figure 5-8 shows the difference between these three styles, using a block as an example.

Orthographic Drawings. The orthographic drawing is used most often in construction. Basically, the orthographic drawing shows the actual arrangement and views of particular objects. The exact form and size of each side of an object is shown in two or more planes, usually at right angles to each other, in an orthographic drawing.

In Figure 5-9, the same cube in Figure 5-8 is used again. However, Figure 5-8 does not show the actual shape of the surfaces. Also, there are no dimensions listed that are necessary when constructing a building. In Figure 5-9, all sides of the cube are shown, so the surfaces are known on all sides. The surfaces shown and the actual dimensions of the sides allow the contractor to construct the building properly.

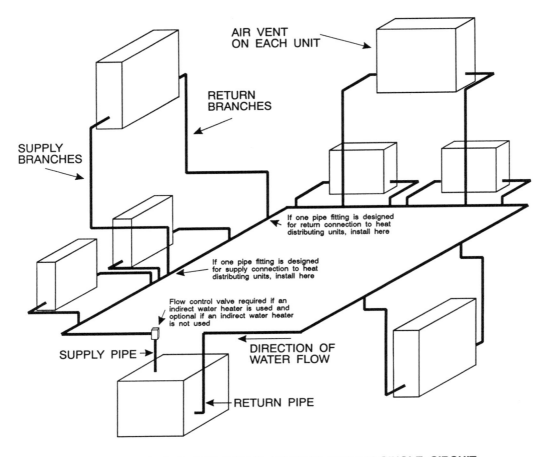

AIR VENT
ON EACH UNIT

RETURN
BRANCHES

SUPPLY
BRANCHES

If one pipe fitting is designed
for return connection to heat
distributing units, install here

If one pipe fitting is designed
for supply connection to heat
distributing units, install here

Flow control valve required if an
indirect water heater is used and
optional if an indirect water heater
is not used

SUPPLY PIPE →

DIRECTION OF
WATER FLOW

← RETURN PIPE

ONE PIPE FORCED HOT WATER HEATING SYSTEM-SINGLE CIRCUIT

Figure 5-6. Plumbing Blueprint in Isometric Style

When creating an orthographic drawing, it is necessary to think of the building surrounded by transparent planes, as in Figure 5-10. Then, these planes are moved to surround the front plane, as in Figure 5-11. Each of the other planes is placed according to where it is in relation to the front plane. If the drawings were all placed back together, they would form the building.

In an actual blueprint on a construction site, all six views of a building are rarely used. Instead, the four sides of the building are shown, the right, left, rear and front elevations. The top view of the building is then replaced with a view from overhead of the floor of the building, called the floor plan. The roof plan is then developed from a view of the top of the building, but a drawing of the bottom of the structure is rarely developed.

Diagrams and Schematics. Diagrams and schematics are used almost exclusively for the mechanical and electrical plans. Symbols are used in both types of drawings. Mechanical plans require several diagrams, including drains and water supply piping; air ducts, grilles, etc.; elevation diagrams of the drainage system; schematic wiring diagrams of the control circuits; and diagrams illustrating pneumatic controls. Electrical plans require a diagram illustrating the circuit layout in a building, a schematic wiring diagram, and an elevation diagram of the power supply.

A diagram is used to simplify components and their related connections. Diagrams are seldom drawn to scale and often use symbols to represent the different components. Lines are then drawn from symbol to symbol. Figure 5-12 shows a HVAC diagram.

Schematic diagrams are actual representations of the electronic controls for HVAC systems, alarm systems, etc. A schematic shows the entire electronic

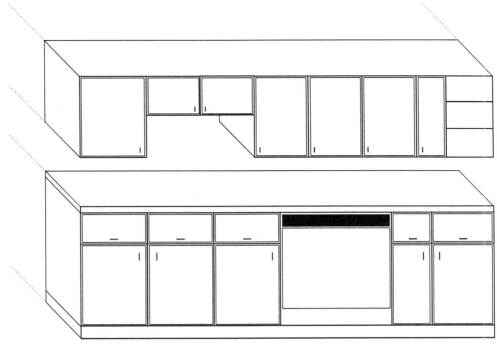

Figure 5-7. Oblique Drawing of Cabinets

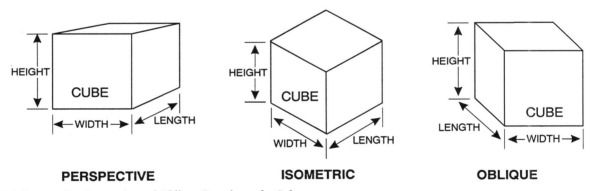

PERSPECTIVE   **ISOMETRIC**   **OBLIQUE**

Figure 5-8. Perspective, Isometric, and Oblique Drawings of a Cube

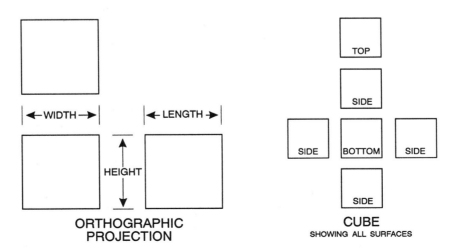

ORTHOGRAPHIC
PROJECTION

CUBE
SHOWING ALL SURFACES

Figure 5-9. Orthographic Drawings of a Cube

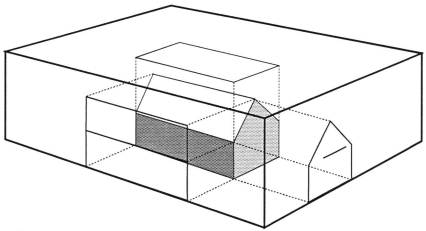

Figure 5-10. Transparent Planes Surrounding the Building

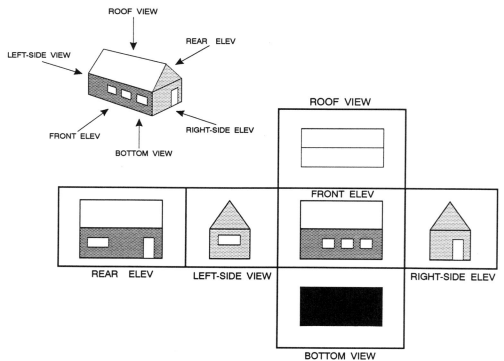

Figure 5-11. Orthographic Drawing of a House

circuit, using symbols, straight lines and notations to represent the electrical components. The schematic specifically shows how to install and troubleshoot a system. Figure 5-13 is an example of a schematic diagram.

The schematic diagram does not attempt to show how the circuit actually looks; therefore, the schematic is not drawn to scale, and the symbols rarely look like the components they are portraying. Instead, the schematic shows the relationship of one part of the

circuit to the rest of the circuit. The straight lines drawn on a schematic diagram should always be horizontal or vertical, never diagonal.

A schematic diagram should always be a closed circuit, and each component represented should be incorporated into this closed circuit. A source of electric current is also always necessary on the diagram, as is a path leading from the source to the component and a path leading from the component back to the source.

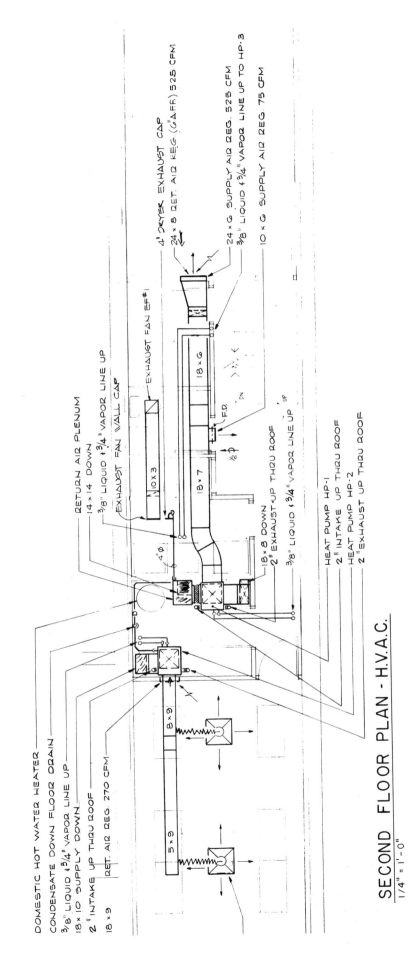

DOMESTIC HOT WATER HEATER
CONDENSATE DOWN FLOOR DRAIN
3/8" LIQUID & 3/4" VAPOR LINE UP
18 × 10 SUPPLY DOWN
2" INTAKE UP THRU ROOF
18 × 9
RET. AIR REG. 270 CFM

RETURN AIR PLENUM
14 × 14 DOWN
3/8" LIQUID & 3/4" VAPOR LINE UP
EXHAUST FAN WALL CAP

EXHAUST FAN EF#1

4' DRYER EXHAUST CAP
24 × 8 RET. AIR REG. (6"ΔFR.) 525 CFM
24 × 6 SUPPLY AIR REG. 525 CFM
3/8" LIQUID & 3/4" VAPOR LINE UP TO HP-3
10 × 6 SUPPLY AIR REG. 75 CFM

18 × 6

18 × 7

F.D.

18 × 8 DOWN
2" EXHAUST UP THRU ROOF
3/8" LIQUID & 3/4" VAPOR LINE UP

HEAT PUMP HP-1
2" INTAKE UP THRU ROOF
HEAT PUMP HP-2
2" EXHAUST UP THRU ROOF

8 × 9

5 × 9

SECOND FLOOR PLAN - H.V.A.C.
1/4" = 1'-0"

Figure 5-12. HVAC Diagram

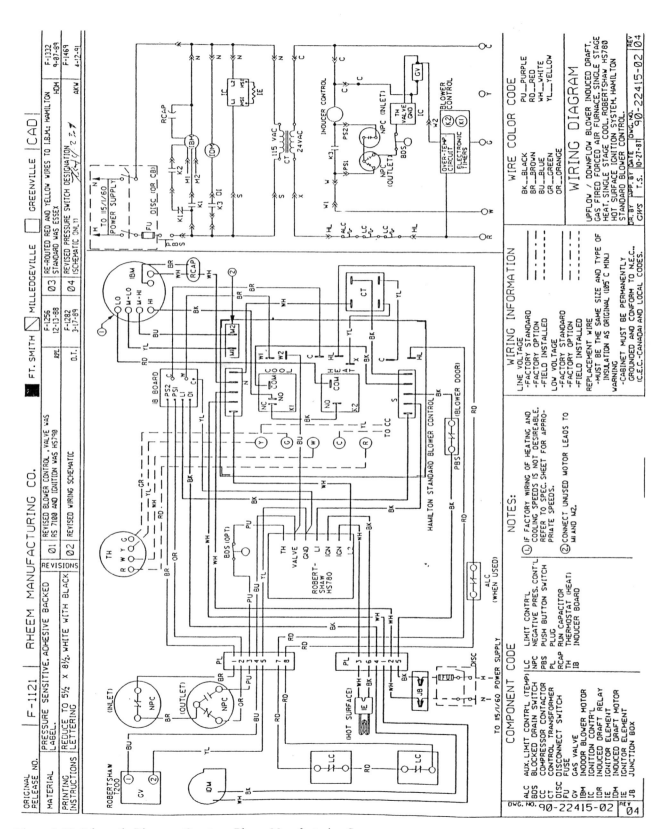

Figure 5-13. Schematic Diagram. Courtesy, Rheem Manufacturing Company.

QUESTIONS FOR CHAPTER 5

Q-1   Name the three different styles of construction drawings.

A-1   _____

_____

_____

Q-2   Which type of pictorial drawing is the most realistic?

A-2   _____

_____

_____

Q-3   Which type of pictorial drawing is preferred?

A-3   _____

_____

_____

Q-4   What is an orthographic drawing?

A-4   _____

_____

_____

Q-5   Which diagrams does a mechanical plan require?

A-5   _____

_____

_____

Q-6   What is a diagram?

A-6   _____

_____

_____

Q-7   What is a schematic diagram, and when is it used?

A-7   _____

_____

_____

Q-8   Are diagrams drawn to scale?

A-8   _____

_____

_____

Q-9   What is an oblique drawing?

A-9   _____

_____

_____

Q-10 What is an isometric drawing?

A-10 _____

_____

_____

CHAPTER 6

BLUEPRINT CATEGORIES

Every blueprint designed for building construction is important. This chapter discusses the different blueprints associated with most construction projects.

BLUEPRINT CATEGORIES

The complete set of blueprints, called working drawings, is usually comprised of eight categories. These categories are: (1) site plans, (2) foundation plans, (3) floor plans, (4) elevation drawings (5) sectional drawings, (6) structural framing plans, (7) mechanical plans, and (8) electrical plans.

Only the general contractor, who is responsible for the total building construction, has a complete set of working blueprints. The subcontractors working under the general contractor have only the blueprints that apply to their specific type of work.

Site Plans. The site plan shows a view of the building on the property as viewed from overhead. This plan is drawn to scale and details the sidewalks, trees and any other existing features. Figure 6-1 is an example of this type of site plan.

A certified land surveyor typically prepares the initial site plan, based on the property deed. This initial plan shows only the lengths of the property boundaries. Sometimes a complete field study is performed in addition to the property survey, and this includes land characteristics, such as whether the land is wooded, hilly, etc.

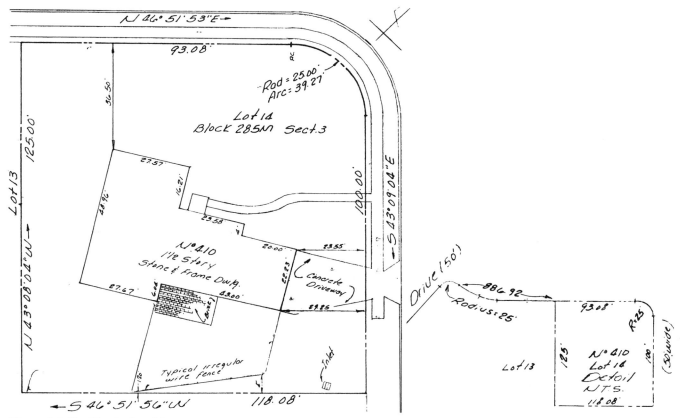

Figure 6-1. Site Plan by Land Surveyor

When an architect draws a site plan, it is normally very detailed, Figure 6-2. This plan usually includes property lines and existing and new contour lines. Water and telephone lines, as well as electrical power lines also normally appear on the site plan. The architect may include precise notes concerning which side of the building faces north, the names of the property owners surrounding the building, monuments, etc. The architect also provides a legend, which is a list and description of symbols appearing on the site plan. The legend is helpful to those who may not otherwise be familiar with a site plan.

Foundation Plans. This type of plan shows the contractor where to locate the footings, foundation walls, sills, columns, and girders.

Footings distribute the weight of the building over a sizable area. The material most often used for footings is concrete, as it can withstand heavy weights and it does not decay. Sometimes steel is used to reinforce the concrete. Footings basically provide the support for the building under the ground line. Figure 6-3 shows several types of footings.

Foundation walls support the weight of the building above the ground. This weight is then transferred to the footings. Foundation walls are made of concrete, however other materials, such as stone or brick can also be used. If the building is constructed with a basement, the foundation walls are also used as basement walls. Figure 6-4 shows the different types of materials used for foundation walls, as well as how they can be used as basement walls.

Sills are normally made of wood, and these are fastened to the foundation with anchor bolts. Their main purpose is to provide a base to which the exterior walls can attach to the foundation. Most local codes require a sill to be installed with some sort of termite protection, as the sill is made of wood and located close to the ground line. Figure 6-5 shows where sills are placed in a building.

Columns are normally made of wood, brick or concrete, and these are used to support the floor. In some types of construction, columns may be the only support of a structure. Or, the columns can be used along with a foundation wall. Figure 6-6 illustrates the different types of columns used in construction.

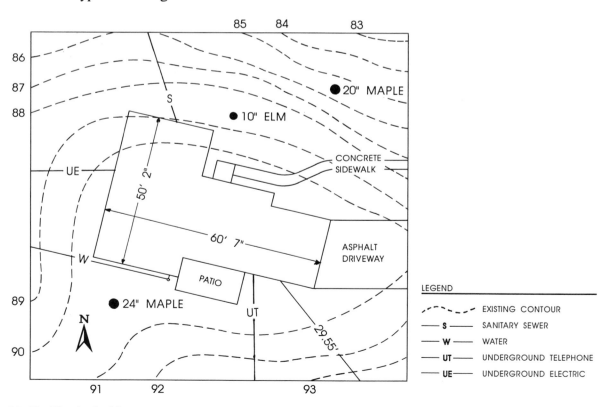

Figure 6-2. Site Plan by Architect

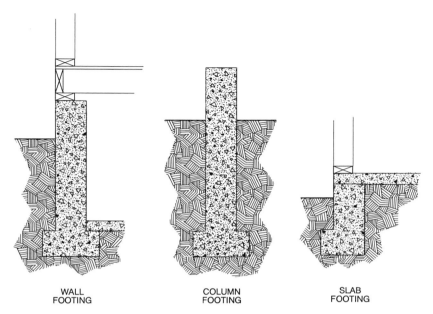

WALL
FOOTING

COLUMN
FOOTING

SLAB
FOOTING

Figure 6-3. Footings

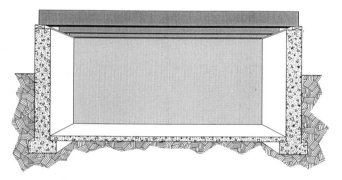

Figure 6-4. Foundation Materials and Walls

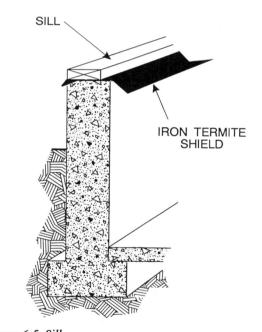

SILL

IRON TERMITE
SHIELD

Figure 6-5. Sill

The girders are horizontal supports that are placed underneath the floor. They rest on columns and are secured to the foundation wall. Girders are typically constructed of wood. Figure 6-7 shows how the girders are installed in a building.

If a building does not have a basement, slab construction is indicated. When this occurs, it may be necessary to place the air conditioning ducts or heating pipes into the concrete slab. However, this is indicated on the HVAC mechanical plan. Figure 6-8 is a representational foundation plan.

Floor Plans. Floor plans are very important blueprints, as they contain the most information concerning the design of the building. Floor plans are basically complete views of each floor as they appear from overhead, therefore, they are considered to be orthographic views. These plans include the location of all doors, windows, walls, etc. Basically, the floor plan covers every item located between the floor and the ceiling. Floor plans are also created to show the HVAC, electrical and plumbing systems. There are two types of floor plans: general-design and working-drawing floor plans.

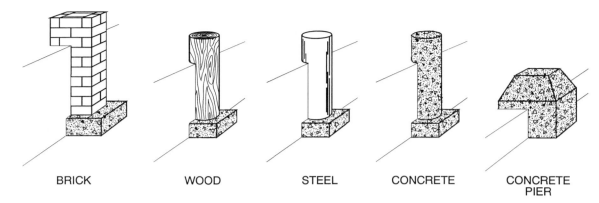

BRICK   WOOD   STEEL   CONCRETE   CONCRETE PIER

Figure 6-6. Columns

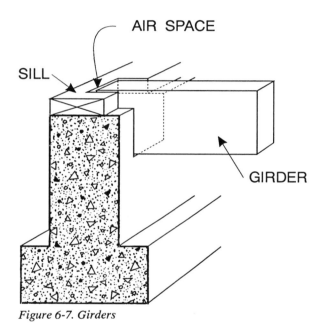

Figure 6-7. Girders

The general-design floor plan shows only the basic layout and arrangement of an area. These plans do not include much detail and are used primarily for sales purposes. There is insufficient information on these floor plans to really be of use in actual construction. General-design floor plans are, however, drawn to scale, and they do include the dimensions for each room. These floor plans are also used to determine the general design of an area. Figure 6-9 shows the most popular general-design floor plan.

The working-drawing floor plan is used primarily for actual construction. The main purpose of these floor plans is to provide the contractor with the information necessary to correctly interpret the architect's design. These floor plans diminish the confusion that may

exist between the architect and the contractor. The working-drawing floor plan usually consists of many separate drawings, including the design of the building, and the wiring and plumbing of the building. Figure 6-10 is an example of a working-drawing floor plan concerning the design of a house. Figure 6-11 shows the plumbing layout in floor plan form.

Elevation Drawings. Elevation drawings are used to depict the exterior or interior walls of a building, and they are most often seen in construction. These drawings are a variation of the orthographic drawing, discussed in the previous chapter. Elevation drawings consist of four separate blueprints, each showing one side of the building; hence, there is a left elevation, right elevation, front elevation, and rear elevation. Figure 6-12 shows one view of an exterior elevation drawing.

On elevation drawings, the architect normally indicates the directions of north, south, east, and west. When these directions are indicated, they are also used to title each elevation. For example, if the front elevation is facing north, that elevation becomes the north elevation. The rear elevation becomes the south elevation, and the left and right elevations become the east and west elevations, respectively.

Floor plans represent the horizontal arrangement of a building interior and exterior elevations show the exterior walls of the building, so interior elevations are needed to depict the interior design of the walls and their components in the vertical plane. Often, interior elevation drawings are used for kitchen and bathroom walls to indicate cabinet and soffit height and depth. Interior elevation drawings are sometimes

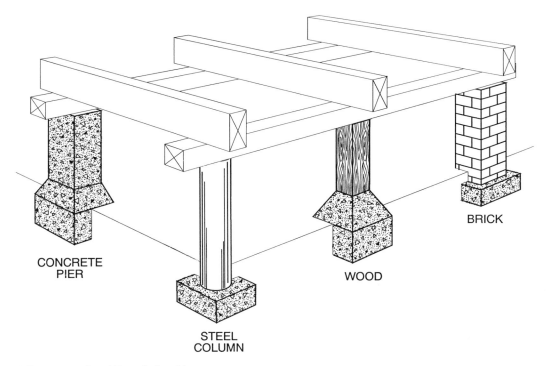

Figure 6-8. Representational Foundation Plan

CONCRETE PIER

STEEL COLUMN

WOOD

BRICK

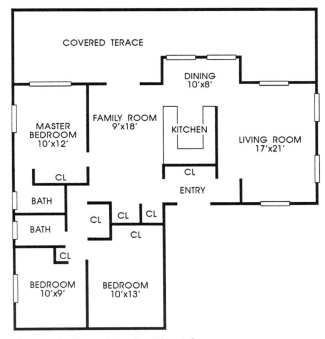

Figure 6-9. General-Design Floor Plan

prepared with the floor line on the bottom of the drawing. Figure 6-13, however, is an example of an interior elevation drawing of a window.

Sectional Drawings. When constructing a building, the contractor needs to know about the internal construction of various elements in the building. If no sectionals are provided, then dotted lines have to be drawn in to represent the internal workings of the building. These dotted lines are not very clear on paper, and they have a tendency to be confusing. Therefore, sectional drawings are used. Figure 6-14 is an example of the dotted-line method.

There is a big difference between an interior elevation drawing and a sectional or detailed drawing. Interior elevation drawings detail all or part of the four walls in a particular room (i.e., kitchen or bathroom). The sectional drawing usually details just part of a wall or a specific section of the building. This is illustrated by Figures 6-13 and 6-15. Figure 6-13 showed the interior elevation of a window. Figure 6-15 shows the sections or details of Figure 6-13. Notice that in Figure 6-13 the sections are called out, and these are shown in Figure 6-15.

When creating a sectional drawing, the architect imagines a plane cutting the building into two sections. This plane is called the cutting plane. The resulting drawing represents what the architect would

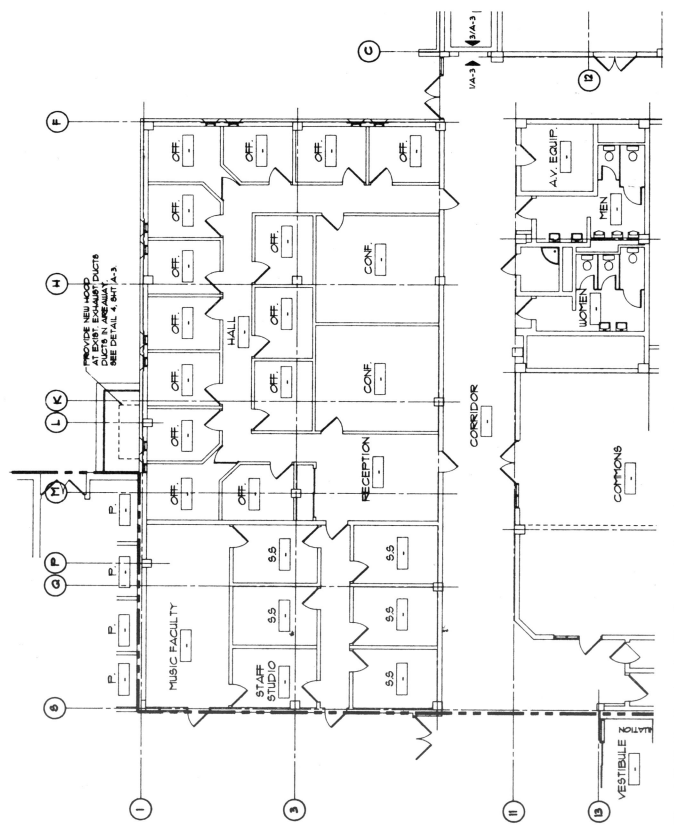

Figure 6-10. Working-Drawing Floor Plan. Courtesy, Northwest Blueprint and Supply Company.

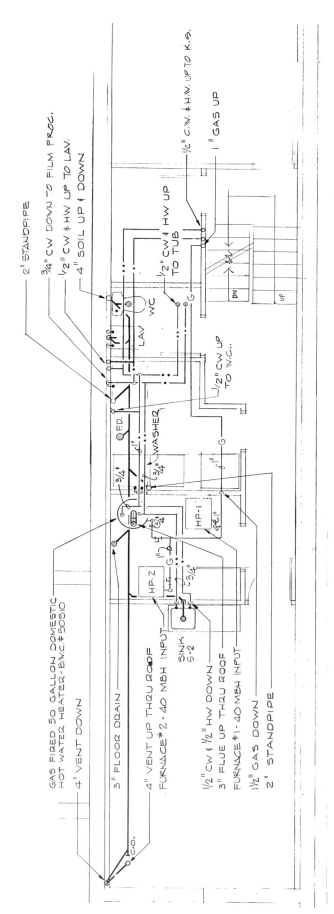

SECOND FLOOR PLAN - PLUMBING

1/4" = 1'-0"

Figure 6-11. Plumbing Layout in Floor Plan Form

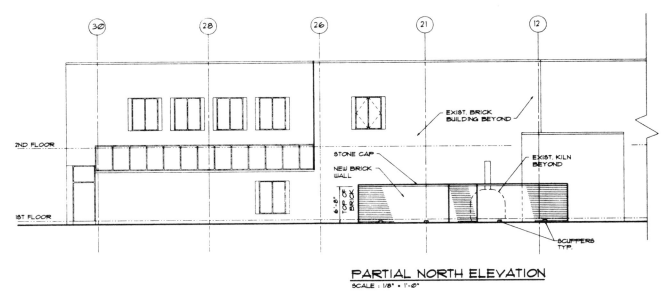

PARTIAL NORTH ELEVATION
SCALE : 1/8" = 1'-0"

Figure 6-12. Exterior Elevation Drawing. Courtesy, Northwest Blueprint and Supply Company.

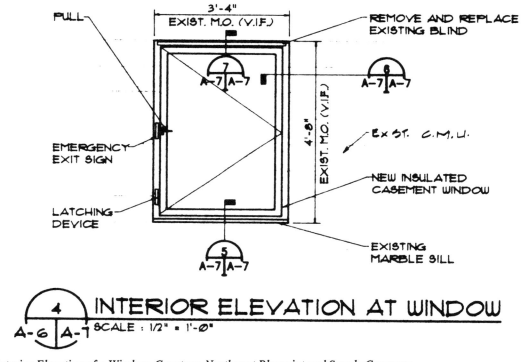

INTERIOR ELEVATION AT WINDOW
SCALE : 1/2" = 1'-0"

Figure 6-13. Interior Elevation of a Window. Courtesy, Northwest Blueprint and Supply Company.

see if part of the building were gone. The architect can use this cutting plane to create just about any view of the building. For example, the architect may want to "cut" part of the building at a 90-degree angle, or "cut" straight through the middle of the building. There are many different options an architect can use when creating a sectional drawing. It is interesting to note that a floor plan is really a horizontal sectional drawing.

Occasionally, the entire building is used in the sectional drawing, and this is called a full section. Normally, however, specific elements of the building require a sectional, and these are called detailed sections. Figure 6-16 is an example of a detailed wall section.

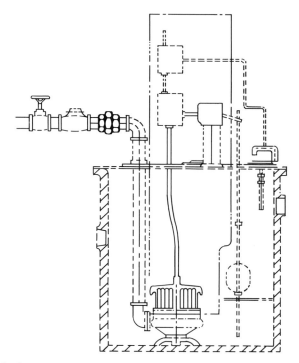

Figure 6-14. Dotted-Line Method

Structural Framing Plans. These plans show the many elements needed (i.e., walls, columns, beams, trusses, etc.) to provide a building with strength and support. No matter which materials or methods are employed, the principles used in structural design do not change from one building to the next. These basic principles are a roof, which is supported by the walls. These walls, in turn, are supported by the foundation, which is supported by footings.

Every building requires many different types of framing plans. For example, there is a need for floor framing plans, wall framing plans, and roof framing plans. Figure 6-17 shows how all these framing plans fit together when constructing a building.

Framing plans are based on the type of weight each element will carry once the building is complete. This weight can be classified into two categories: live weight and dead weight. Live weight varies from building to building, depending on where the building is located, and how the building is used. Wind, snow, and rain are all considered live weights, as are people and furnishings in the building. Dead weight consists of the weight of the materials used to construct the building. For example, the dead weight of the roof rests on the walls, and the dead weight of the walls rests on the foundation.

Mechanical Plans. Mechanical plans usually include the information necessary to install the heating, ventilating, air conditioning and plumbing systems. These plans include the complete layouts of the systems, as well as other drawings, such as floor plans and sectional drawings, in order to fully illustrate the systems. Basically, the mechanical plans cover the areas that deal with the comfort level of the people occupying the building.

Electrical Plans. The architect does not actually create the electrical plans; instead, the architect hires an engineer to design the electrical system. Like the mechanical plans, these plans are normally comprised of several other types of plans already discussed. For example, electrical plans usually consist of a site plan, floor plan, and elevation plan, as well as the schematic wiring diagrams. In order to read an electrical blueprint, it is necessary to have a working knowledge of electrical circuits. Also, electrical symbols are often used on these blueprints, so it is necessary to recognize and understand what these symbols mean.

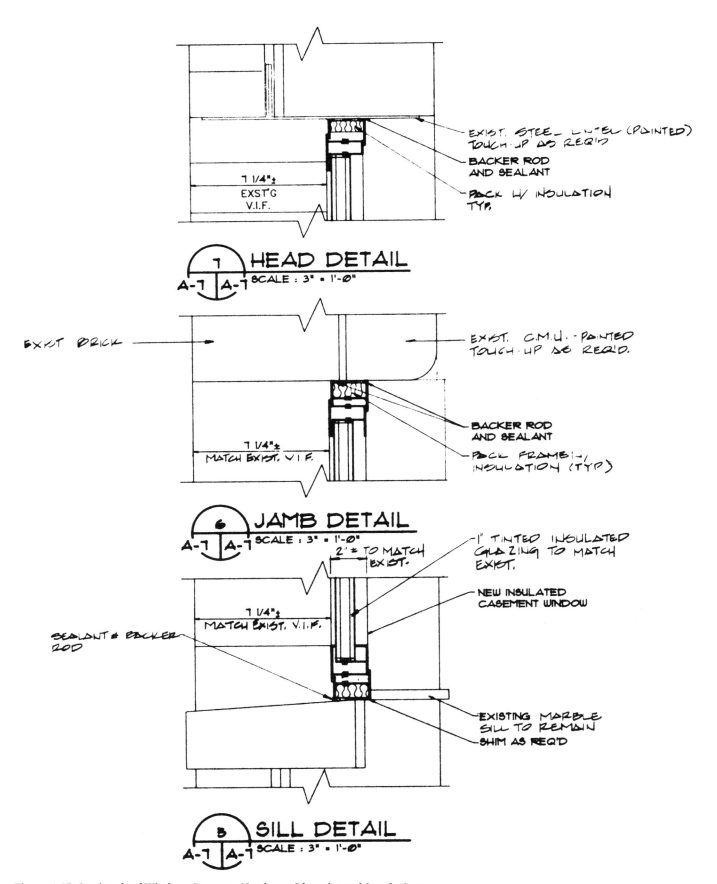

Figure 6-15. Sectionals of Window. Courtesy, Northwest Blueprint and Supply Company.

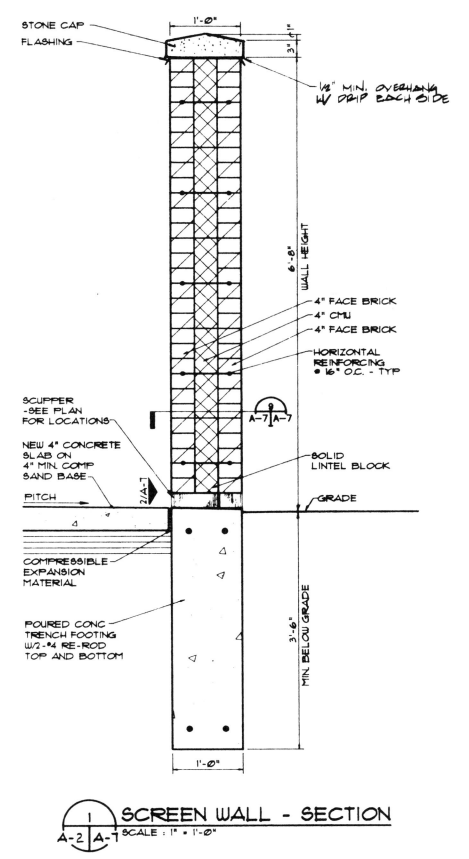

STONE CAP

FLASHING

1'-0"

3" 1"

1/2" MIN. OVERHANG W/ DRIP EACH SIDE

WALL HEIGHT

6'-8"

4" FACE BRICK
4" CMU
4" FACE BRICK

HORIZONTAL REINFORCING @ 16" O.C. - TYP

SCUPPER -SEE PLAN FOR LOCATIONS

A-7 A-7

NEW 4" CONCRETE SLAB ON 4" MIN. COMP SAND BASE

SOLID LINTEL BLOCK

2/A-7

PITCH

GRADE

COMPRESSIBLE EXPANSION MATERIAL

POURED CONC TRENCH FOOTING W/2 -#4 RE-ROD TOP AND BOTTOM

MIN. BELOW GRADE

3'-6"

1'-0"

SCREEN WALL - SECTION

A-2 A-7

SCALE : 1" = 1'-0"

Figure 6-16. Detailed Wall Section. Courtesy, Northwest Blueprint and Supply Company.

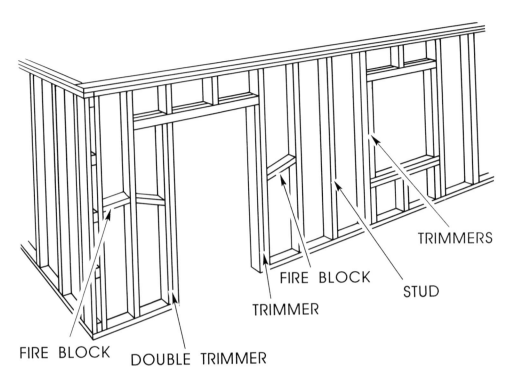

Figure 6-17. Framing a Building

QUESTIONS FOR CHAPTER 6

Q-1   What type of information is on a site plan?

A-1   _____

      _____

      _____

Q-2   Who normally draws the site plan?

A-2   _____

      _____

      _____

Q-3   What do the following mean:

A-3   footings -

      sills -

Q-4   Which plan gives the most information concerning the design of a building?

A-4   _____

      _____

      _____

Q-5 What is the main goal of the working-drawing floor plan?

A-5 _____

_____

_____

Q-6 What is the purpose of a sectional drawing?

Q-6 _____

_____

_____

Q-7 What is a cutting plane?

Q-7 _____

_____

_____

Q-8 In a sectional drawing, why are dotted-lines not acceptable?

A-8 _____

_____

_____

Q-9 In a structural framing plan, what is dead weight and why is it important?

A-9 _____

_____

_____

Q-10 What are mechanical plans?

A-10 _____

_____

_____

CHAPTER 7

READING BLUEPRINTS

In a previous chapter, the eight different categories of blueprints were discussed in general terms. In this chapter, each of these categories will be looked at more closely, in terms of actually reading these particular blueprints.

READING SITE PLANS

Site plans show the building placement in relationship to the property. Sidewalks, trees, and any other features on the property are also included on the site plan.

In order to read site plans correctly, the following points should be considered (Figure 7-1 details some of these points):

• Only the outline of the building is shown - the interior walls are rarely, if ever, detailed.

• As most building codes require a building to be located a certain distance away from the property lines, property lines must detail the legal limits of the property on all sides.

• The overall size of the building is shown.

• The distance from the outside wall of each building to at least two property lines and the direction of each property line is usually included.

• If there is more than one building on the property, each building is shown in relation to a main building or to the property lines.

• All driveway positions and sizes are shown, as are the sizes and locations of all sidewalks.

• The engineer's scale is normally used on site plans.

• An arrow showing the northern direction is always included.

• Streets bordering the property are labeled.

• The grade elevations of patios, decks, driveways, etc., are shown.

• Dimension lines are placed either outside the property line or actually on the property line.

• Symbols usually represent the materials used for patios, decks, driveways, etc.

• The relationship of the utility lines to the property and the building is shown, as are the functions of the utility lines.

• All bodies of waters are labeled and outlined.

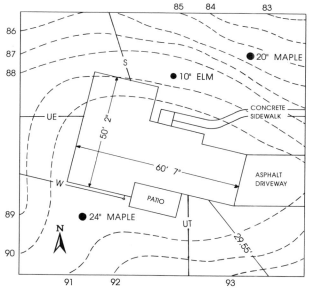

Figure 7-1. Site Plan Details

READING FOUNDATION PLANS

Every building constructed requires a foundation. This foundation is actually a support system for the entire building. In addition, the foundation can also provide waterproof walls for a basement.

Before proceeding further, it is necessary to state that there are three different kinds of foundations: the slab foundation, the T foundation, and the column foundation. The type of foundation used depends on the weight and size of the building, as well as the climate, soil, and building codes. Figure 7-2 shows these different foundation types.

The following points should be considered when reading foundation and basement plans:

- Footings, wall supports, and other support devices are normally outlined.

- Pictorial drawings normally show each portion of the foundation plan.

- Porches, patios, and other external concrete slabs are detailed.

- Symbols represent the different types of materials used in foundations.

- Complete basement plans show laundry and bathroom fixtures, floor drains, fireplaces, and the locations of posts and girders.

- The direction of floor joists is shown.

- Basements that are excavated or partially finished are always labeled as such.

- When a basement contains a fireplace, elevation views, sectional views, and floor plan views are included.

READING FLOOR PLANS

Floor plans contain the most information concerning the design of the building. Contractors often read

floor plan dimensions incorrectly, therefore, it is necessary to read these floor plans carefully. The following suggestions should be considered when reading floor plans.

- If a building contains more than one story, each floor plan will be labeled as to which floor it is.

- Abbreviated floor plans only show the overall dimensions of rooms.

- If dimension lines are not present in a room, the width and length are indicated.

- Dimensions that are less than 1 foot are expressed in inches, while dimensions greater than 1 foot are shown in feet and inches.

- Window and door dimensions are often placed directly on the window or door symbol.

- When a building component is too small for dimension numerals, the dimension is positioned outside the extension lines. Or, the arrowheads may be placed outside the extension lines.

- Despite the drawing scale, dimensions always represent the actual size of the building.

- Notes on the floor plan often refer to the sizes and directions of the floor coverings, ceiling joists, beams, etc.

- The wall thickness dimension is the total of each component in the wall and partition (i.e., panel, brick, tile, etc.).

- Regardless of the wall thickness, room sizes are determined from interior wall to interior wall.

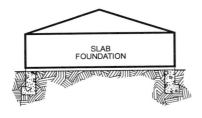
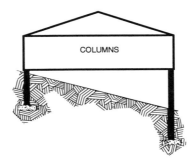
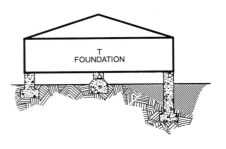

Figure 7-2. Foundation Types

- The walls and partitions of a floor plan should be memorized.

- Major component position (i.e., doors, windows, stairs, etc.) should be studied.

- The dimensions should be examined in their entirety, starting with the outside dimensions and ending with the inside dimensions.

- All notes on the floor plan should be read.

ELEVATION DRAWINGS

Elevation drawings show the exterior or interior walls of a building. These are the most common drawings found in construction. Reading elevation dimensions can be tricky, so the following points should be considered:

- The position of the ground line is calculated as the ground height above the datum line. The datum line is a constant horizontal plane, such as sea level.

- Only vertical distances are measured (Figure 7-3).

- Dimension lines indicate footing depths.

- Window or door styles are sometimes shown directly.

- Window or door heights are calculated from the floor to the top of the window or door.

- The angle of the roof, also called roof pitch, is indicated by a measurement called rise over run, illustrated in Figure 7-4.

- The distance between the floor line and the ground line is always indicated.

- Room heights are determined by the distance between the floor line and the ceiling line.

- Slab thickness is indicated with a dimension line.

SECTIONAL DRAWINGS

As stated previously, sectional drawings depict the internal construction of different elements in the building. There are various types of detailed sectional drawings, and each requires special attention. The following points should be considered when viewing detailed section drawings:

- Symbols are often used in sectional drawings to show material detail.

- Footing width and height are detailed to show the material used and the foundation wall location.

- Removed sections are used when a very large area needs detailed sections. Figure 7-5 details these removed sections.

- Wall sections show the entire internal wall, from the footing to the roof.

- Actual window construction is detailed.

- Jamb sections are shown by way of a horizontal cutting plane through the entire window. As the two jambs are mirror images of one another, only one jamb is normally shown.

- Jamb sections of the doors are used often.

- The method in that the beams support the joists and the walls or columns support the beams is detailed.

- Built-in furniture, fixtures, etc., are shown as detailed sections, as are other uncommon construction methods.

STRUCTURAL FRAMING PLANS

Framing plans show the materials needed to support a building. There are many different types of framing plans, including floor, wall, and roof framing plans. There are then several categories under each of these sections. The points outlined here are designed to be simple, general guidelines to reading framing plans.

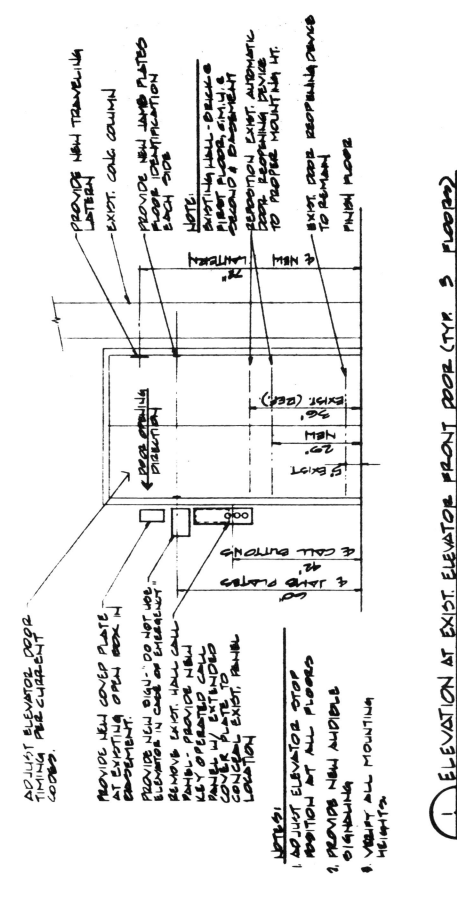

Figure 7-3. *Elevation Drawing Measures Vertical Distances. Courtesy, Northwest Blueprint and Supply Company.*

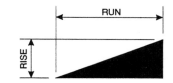

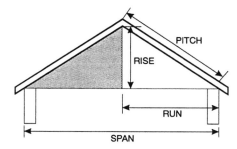

Figure 7-4. Roof Pitch

When reading floor framing blueprints, the following points should be considered:

- Girders can consist of either wood or steel.

- Joists can either be placed from one girder to another or from a girder to the foundation wall.

- Contractors determine the floor framing plan if the floor framing plan is not included in the architectural blueprints.

- Diagonal lines indicate chimney and stair openings.

- Supplemental joists are called headers. Headers are used to support regular joists that have been cut.

- Stair openings show joist and header positions.

- The length, width, and height of headers at stair openings are usually shown.

- Elevation and sectional views of the floor framing plans are frequently included.

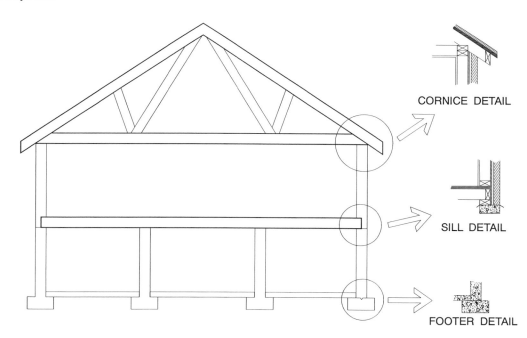

CORNICE DETAIL

SILL DETAIL

FOOTER DETAIL

Figure 7-5. Sectionals of Wall Framing Plan

When reading wall framing plans, the following points should be considered:

- Separate plans are made for exterior and interior walls.

- Window rough openings are shown.

- Stud spacing, width, and height are detailed.

- Elevation drawings are often used to illustrate wall detail.

- The stud layout can either be shown as a complete plan or as a detail plan. The complete plan shows all the studs in the wall framing plan, and the detail plan shows just a few studs at critical points, usually at intersections.

- Inside and outside corners are normally shown as sectional drawings.

When reading roof framing plans, the following points should be considered:

- A roof framing plan shows the exact position of each item located in the roof.

- Roof pitch is detailed.

- Chimney intersections, skylights, and vent pipes are all shown.

- A roof plan only shows an overhead view of the roof, without all the details shown in a roof framing plan.

MECHANICAL PLANS

Mechanical plans include information concerning heating, ventilating, air conditioning and plumbing. While these plans may seem somewhat interrelated, separate HVAC and plumbing blueprints are almost always provided.

When reading HVAC duct system blueprints, the following points should be considered:

- Duct systems can be drawn as floor plan, elevation, or sectional views.

- Symbols representing the various heating devices are shown.

- Duct systems vary, depending on the type of heat used in the building (i.e., forced air, hot water, steam, etc.).

- When a brief heating and/or air conditioning plan is used for a forced-air system, only the warm air outlets are shown.

- When only a brief heating and/or air conditioning plan is supplied in a forced-air system, the contractor must determine the duct system requirements.

- Forced-air duct systems can be laid out as an individual duct system (Figure 7-6), an extended plenum system, a perimeter radial system, or a perimeter loop system.

- Electric heat does not require a duct system, therefore notations concerning electric wires or panels are normally made on floor plans.

- When an electric heating system is used, the power supply and thermostat are shown on electrical plans.

- Hot-water duct systems can be laid out as a series loop system, a radiant system, a one-pipe system (Figure 7-7), or a two-pipe system.

- Steam duct systems in blueprints are shown in the same manner as the hot-water system.

- Heat pump duct systems use the same layout as in the forced-air system.

- When an air conditioner is installed with a forced-air heating system, the same blower and vents are normally used, so the blueprint is essentially the same as for a forced-air heating system (Figure 7-8).

- Duct sizes, return and supply register sizes and the cubic feet of air per minute (cfm) each register delivers are normally detailed on the blueprint.

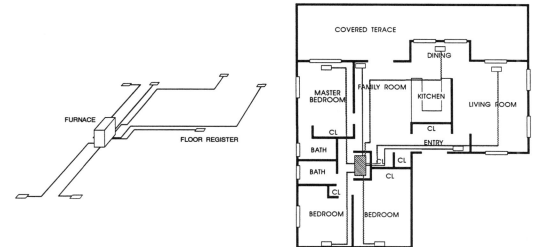

Figure 7-6. Individual Duct System

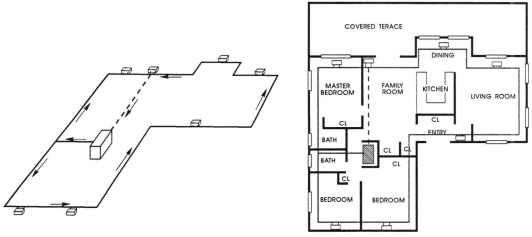

Figure 7-7. One-Pipe System

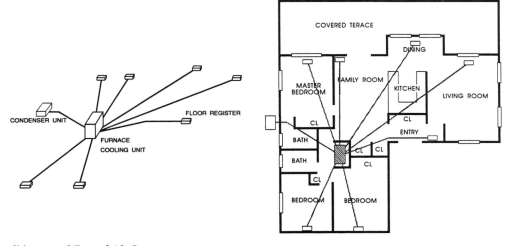

Figure 7-8. Air Conditioner and Forced-Air System

• Figures 7-9 and 7-10 show two different methods used to lay out a duct system floor plan view. Figure 7-9 shows the approximate shapes and sizes of the duct and registers. Figure 7-10 shows the duct as one size throughout the drawing. Care should be taken when reading this latter drawing.

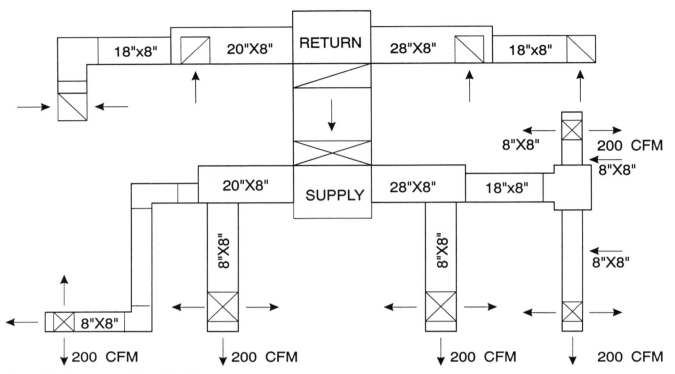

Figure 7-9. Duct System Floor Plan View

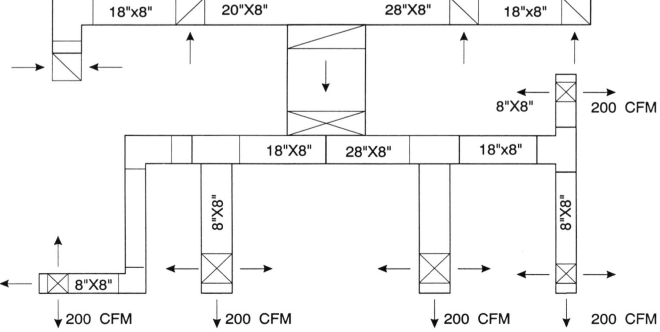

Figure 7-10. One-Size Duct System Floor Plan View

- Figure 7-11 shows another duct system floor plan view, except this one uses a single line to represent the duct. Arrows show the direction of air flow.

- Figure 7-12 shows the duct system laid out to scale in an actual building.

When reading HVAC piping blueprints, the following points should be considered:

- The different piping in an HVAC plan can include hot water supply and return lines to and from the heater boilers, natural gas or oil supply lines, refrigerant lines, condensate lines, as well as other kinds of piping.

- Piping is normally shown in a floor plan view or isometric diagram.

- Figure 7-13 shows a detailed section of a hot water boiler. This view shows some of the accessories needed in this kind of system.

When reading plumbing blueprints, the following points should be considered:

- The different piping shown includes supply and return water lines.

- Plumbing blueprints are normally either an elevation or floor plan view.

- Toilets, tubs, sinks, vents, traps, and sewer lines are indicated.

- As insulation must be placed around hot water lines, the type and thickness of insulation is indicated.

- If a septic tank is needed, a separate drawing showing only the septic tank is usually provided.

- A floor plan view shows both the water supply and sewage disposal systems; however, an elevation view shows only one system.

ELECTRICAL PLANS

Electrical blueprints relay the information necessary concerning all the electricity in the building. It is necessary to have a working knowledge of electrical circuits in order to read electrical blueprints. When reading electrical blueprints, the following points should be considered:

- Symbols represent outlets and controls.

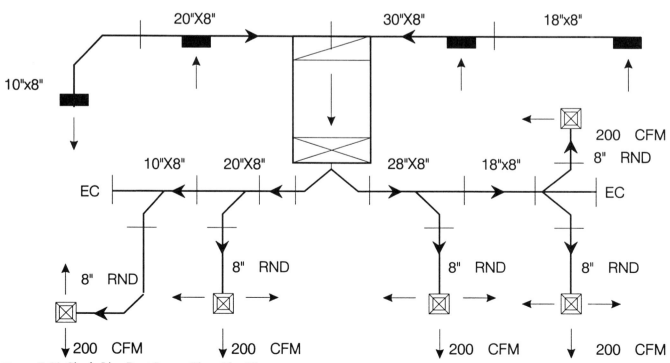

Figure 7-11. Single Line Duct System Floor Plan View

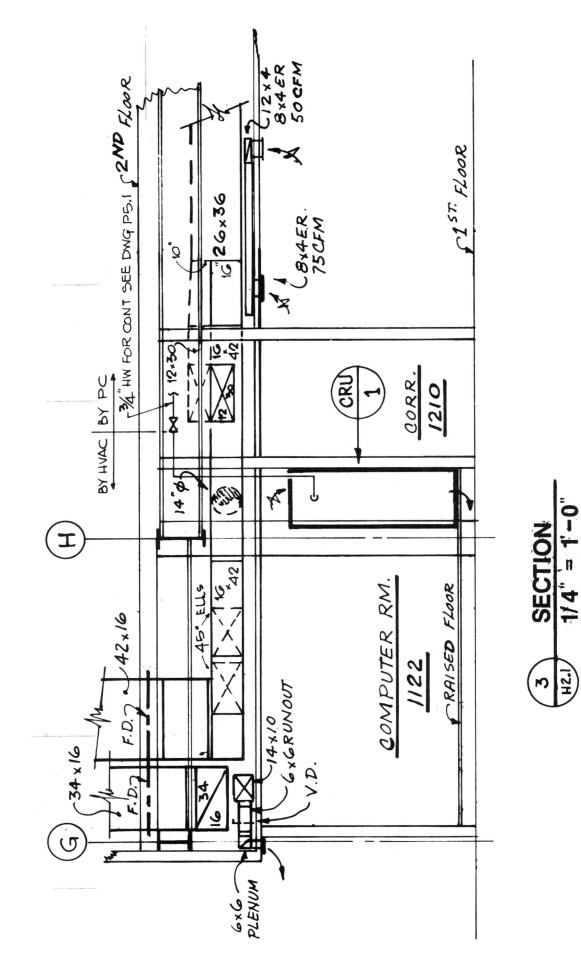

Figure 7-12. Duct System to Scale

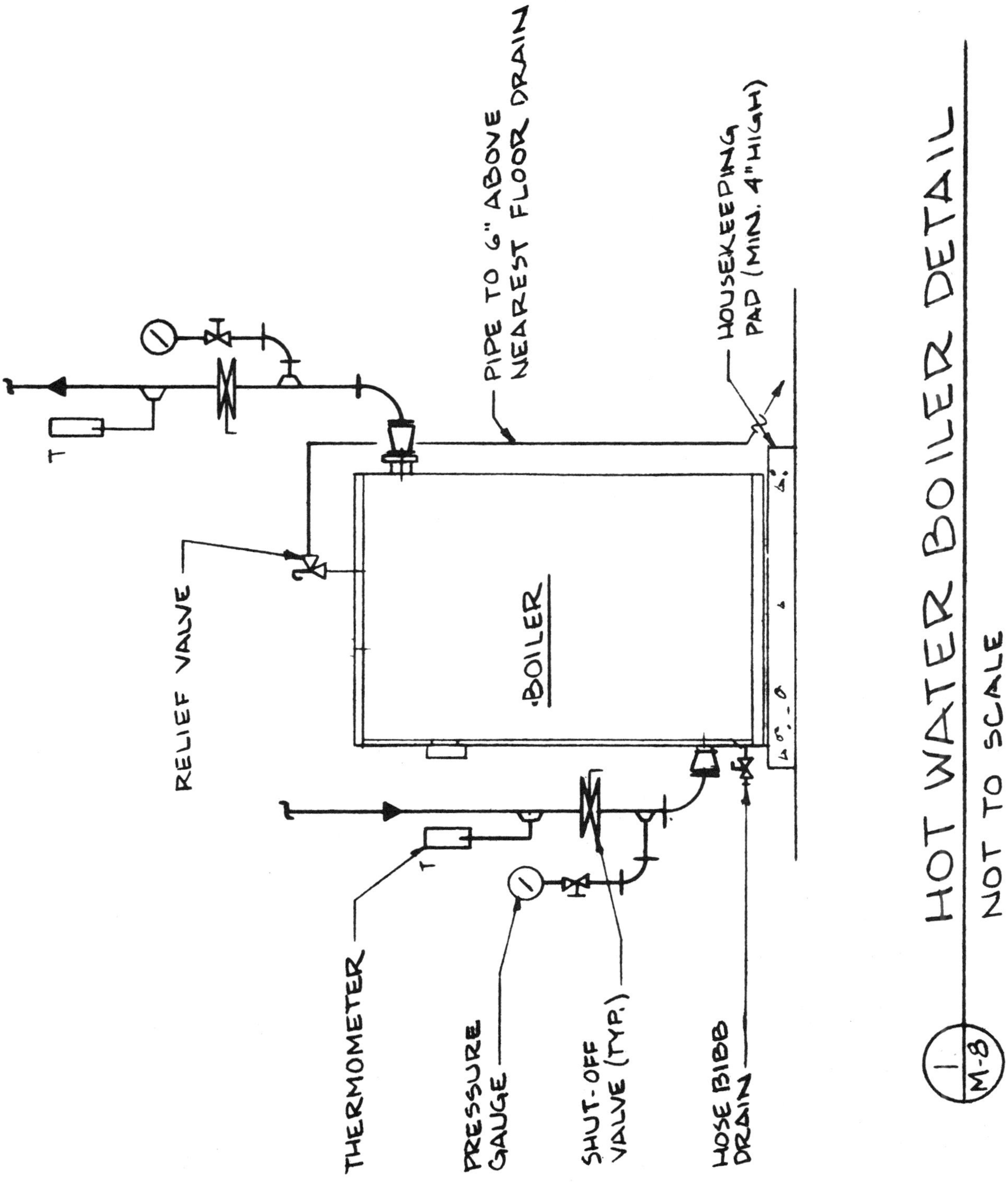

PIPE TO 6" ABOVE
NEAREST FLOOR DRAIN

HOUSEKEEPING
PAD (MIN. 4" HIGH)

RELIEF VALVE

BOILER

THERMOMETER

PRESSURE
GAUGE

SHUT-OFF
VALVE (TYP.)

HOSE BIBB
DRAIN

HOT WATER BOILER DETAIL

NOT TO SCALE

M-8

Figure 7-13. Hot Water Boiler

- Complete wiring plans show the positions of all switches, fixtures, and outlets. Figure 7-14 illustrates a wiring plan.

- Dotted lines connect each fixture to its specific switch. These dotted lines are not the paths of the actual wires.

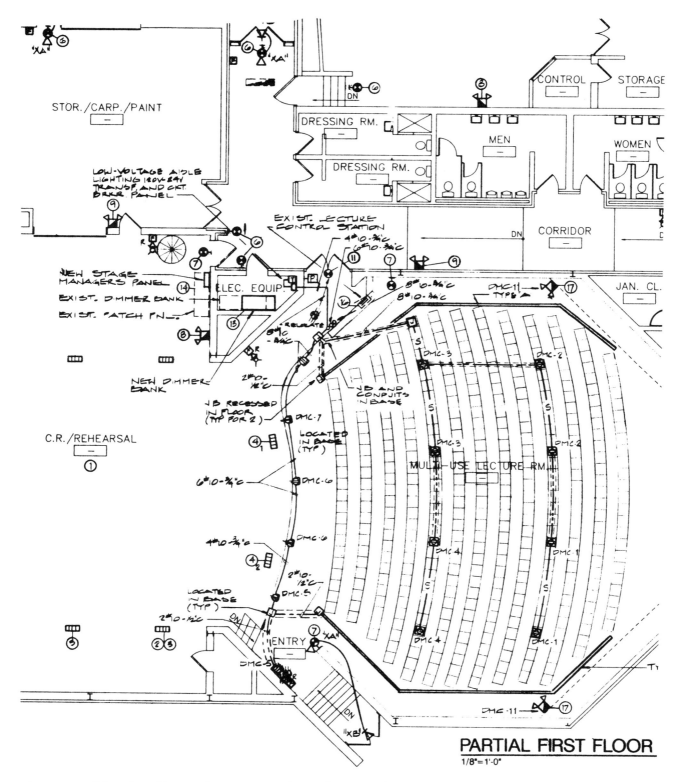

PARTIAL FIRST FLOOR
1/8"=1'-0"

Figure 7-14. Wiring Plan with Legend. Courtesy, Northwest Blueprint and Supply Company.

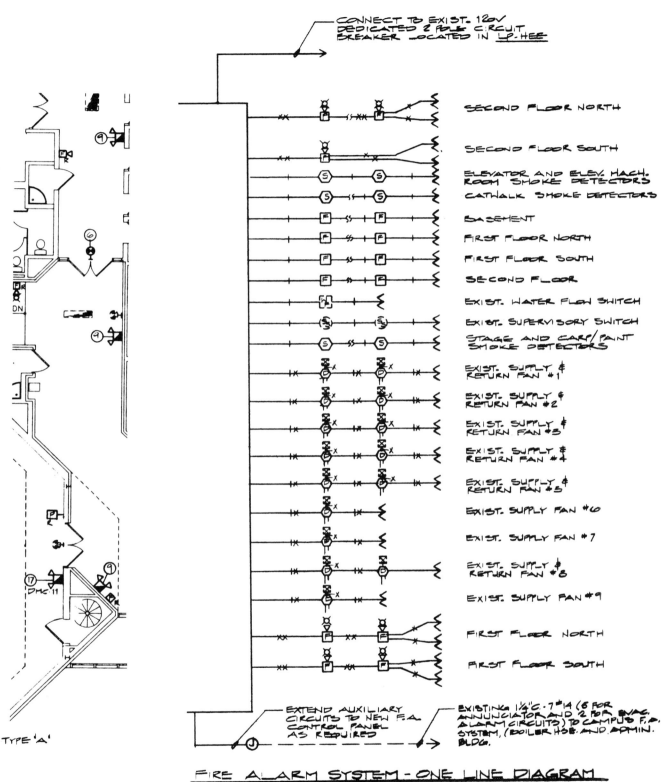

CONNECT TO EXIST. 120V
DEDICATED 2 POLE CIRCUIT
BREAKER LOCATED IN LP-HEE

SECOND FLOOR NORTH

SECOND FLOOR SOUTH

ELEVATOR AND ELEV. MACH.
ROOM SMOKE DETECTORS

CATWALK SMOKE DETECTORS

BASEMENT

FIRST FLOOR NORTH

FIRST FLOOR SOUTH

SECOND FLOOR

EXIST. WATER FLOW SWITCH

EXIST. SUPERVISORY SWITCH

STAGE AND CARP/PAINT
SMOKE DETECTORS

EXIST. SUPPLY &
RETURN FAN #1

EXIST. SUPPLY &
RETURN FAN #2

EXIST. SUPPLY &
RETURN FAN #3

EXIST. SUPPLY &
RETURN FAN #4

EXIST. SUPPLY &
RETURN FAN #5

EXIST. SUPPLY FAN #6

EXIST. SUPPLY FAN #7

EXIST. SUPPLY &
RETURN FAN #8

EXIST. SUPPLY FAN #9

FIRST FLOOR NORTH

FIRST FLOOR SOUTH

EXTEND AUXILIARY
CIRCUITS TO NEW F.A.
CONTROL PANEL
AS REQUIRED

EXISTING 1/4"C. 7 #14 (5 FOR
ANNUNCIATOR AND 2 FOR ENAC.
ALARM CIRCUITS) TO CAMPUS F.A.
SYSTEM, (BOILER HSE. AND ADMIN.
BLDG.

FIRE ALARM SYSTEM - ONE LINE DIAGRAM
NO SCALE

TYPE 'A'

DN

DMC-11

QUESTIONS FOR CHAPTER 7

Q-1   What type of scale is normally used for site plans?

A-1   _____

Q-2   Where are dimension lines placed on a site plan?

A-2   _____

       _____

       _____

Q-3   What are the three different foundation types?

A-3   _____

       _____

       _____

Q-4   What do complete basement plans show?

A-4   _____

       _____

       _____

Q-5   In a floor plan, when dimension numerals cannot fit in a small building component, where are the
       dimension numerals placed?

A-5   _____

Q-6 How does wall thickness affect room dimensions in a floor plan?

A-6 _____

_____

_____

Q-7 What is a datum line?

A-7 _____

_____

_____

Q-8 When discussing roof pitch, what does rise over run mean?

A-8 _____

_____

_____

Q-9 What is a jamb section?

A-9 _____

_____

_____

Q-10 Name the three different framing plans.

A-10 _____

_____

_____

Q-11 Describe a stud detail plan.

A-11 _____

_____

_____

Q-12 Describe the electric heating duct system.

A-12 _____

_____

_____

Q-13 What are some of the different kinds of piping found in an HVAC system?

A-13 _____

_____

_____

Q-14 What types of components are found on plumbing blueprints?

A-14 _____

_____

_____

Q-15 What do dotted lines represent on an electrical blueprint?

A-15 _____

_____

_____

CHAPTER 8
CONSTRUCTION SPECIFICATIONS

Construction specifications are the written guidelines outlining the work and duties of the engineer, architect, contractor, and owner. These specifications and the working blueprints are normally included in the legal contracts signed by all parties.

CONSTRUCTION SPECIFICATIONS

Construction specifications accompany the working set of blueprints. These specifications are of particular importance to the contractor, as they determine the types of materials that are to be used when constructing the building, as well as the required construction methods. Also outlined in the specifications are the installation requirements for the building and the time frame in which the contractor is to have the building completed. The specifications are very detailed, so that no misunderstandings will occur.

A term used often in construction specifications is *same or equal quality*. This means that if a material or component appearing on the construction specifications is not available at the time of construction, the contractor can authorize a substitute, as long as it is of equal quality to the original specification. Often, there is a dispute between the contractor and architect as to the definition of equal quality. For this reason, the contractor should consult with the architect or engineer for written approval before substituting any item.

It is always the responsibility of the contractor to check the blueprint and construction specifications for any discrepancies that may exist. There are some cases when the construction specifications and the blueprints may not agree. If any changes or corrections are to be made, they must be made to both documents. However, before any changes are made, all parties involved must provide written consent.

Specialists trained in the mechanical and legal aspects of construction normally create the specifications. Once the specifications are drawn up, a lawyer usually checks them to ensure there are no legal problems that may occur at a later date.

SPECIFICATIONS OUTLINE

Construction specifications are normally constructed in outline form. The outline is then broken down into sections, and these sections are placed basically in the same order as the actual construction. Each subcontractor normally receives only one of these sections, depending on the specialty. Under each section is a list of materials pertaining to that particular division. Construction specifications vary, depending on the person drawing the specifications, and the type of work to be performed. An example of these specifications is outlined below.

General Information. This section summarizes the work to be done, as well as the materials purchased and the estimated completion date. All drawings and legal documents are included in this section. Every person involved in the project should receive a copy of this section.

Earthwork. Earthwork includes items concerning excavation, proper drainage, filling ground irregularities, and preparing the site for the foundation. Earthwork normally does not include landscaping or excavating for the electrical work.

Concrete. Concrete mixing, placing concrete, porches, patios, sidewalks, and patching are just a few of the items in this category. Typically included in this section is the placement of reinforcement steel and electric conduit.

Carpentry. This section details the types of wood grades required, the nail sizes, and the maximum amount of moisture allowed in the wood. Rough and finish carpentry are both listed in this category, and wood doors are sometimes included here, too.

Roofing. This category covers more than just roofing. Basically, all roofing, waterproofing, and insulation are covered. In addition, siding, sheet metal work, and all sealants are in this category.

Windows, Doors and Glass. The amount of window and door space for each room is calculated, as are the types of glass for the windows, and the proper materials for the doors. Window and door frames, glazing, and weatherstripping are also usually detailed in this section.

Plaster and Lath. The quality of the plaster and lath and the type of workmanship to be used with the plaster and lath are detailed in this division. The instructions for mixing and applying the plaster and lath, as well as the drying time and number of coats are included in this section. Finally, the acoustical requirements, floor treatments, and wall coverings (i.e., drywall) are detailed in this category.

Special Items. This section usually covers fireplaces, flagpoles, shelving, bath and toilet accessories, and any compartments or cubicles.

Masonry. The location and types of stone and brickwork (i.e., fireplaces, chimneys, etc.), as well as the masonry cleaning and mortar are found in this category.

Mechanical. This division covers all aspects of heating, ventilating, air conditioning, and plumbing. Included in this information is the type and size of plumbing lines, the style and manufacturer's name of equipment, the vent pipes, a list of fuels, and the location for heating and air-conditioning units and registers.

Electrical. The electrical outlets, number of circuits, location of telephones, size of wire used, and a detailed list of all electrical components used (including style, manufacturer and model number) are found in this division.

Painting and Miscellaneous Items. Paint color of both the interior and exterior walls is described here, as well as the type of paint to be used, and the instructions concerning the number of coats and the surface preparation. Also included here is any of the finish hardware needed and the location of blacktop surfaces surrounding the building.

OTHER SPECIFICATION INFORMATION

As already stated, most of the construction specifications include extensive information concerning styles, colors, locations, etc. However, the specifications also include guarantees that the specific contractor must provide. These guarantees ensure that the contractor uses the proper, quality materials and installs each item correctly. If a component or material is found to be substandard, then the contractor is held liable. This means the contractor must replace the faulty component or material at the contractor's expense. This is not beneficial to the contractor, so normally, the contractor upholds all parts of the construction specifications.

The specifications normally state that the contractor is responsible for all codes, fees and permits. Therefore, the contractor must be familiar with the local codes required for a specific component, and all fees associated with acquiring a permit, if one is necessary. If the contractor does not find out this information, there could be repercussions at a later date. Figure 8-1 is an example of standard construction specifications.

DESCRIPTION OF MATERIALS

Property address _____ City _____ State _____

Mortgager or Sponsor: Name _____ Address _____

Contractor or Builder: Name _____ Address _____

1) EXCAVATION:
   Bearing soil, type _____

2) FOUNDATION:
   Footings: concrete mix _____ strength psi _____ Reinforcing _____

   Foundation wall: material _____ Reinforcing _____

   Interior foundation wall: material _____ Party foundation wall _____

   Columns: material and sizes _____ Piers: material and reinforcing _____

   Girders: material and sizes _____ Sills: material _____

   Basement entrance areaway _____ Window areaways _____

   Waterproofing _____ Footing drains _____

   Termite protection _____

   Basementless space: ground cover _____ insulation _____ foundation vents _____

   Special foundations _____

   Additional information: _____

3) CHIMNEYS:

   Material _____ Prefabricated (make and size) _____

   Flue lining: material _____ Heater flue size _____ Fireplace flue size _____

   Vents (material and size): gas or oil heater _____ water heater _____

   Additional information: _____

4) FIREPLACES:

   Type: ☐ solid fuel   ☐ gas-burning   ☐ circulator (make and size) _____ Ash dump and clean-out _____

   Fireplace: facing _____ lining _____ hearth _____ mantel _____

   Additional information: _____

Figure 8-1. Construction Specifications

5) EXTERIOR WALLS:

Wod frame: wood grade, species _____ ☐ Corner bracing   Building paper or felt _____

    Sheathing _____ thickness _____ width _____ ☐solid ☐ spaced _____ " o.c. ☐ diagonal _____

    Siding _____ grade _____ type _____ size _____ exposure _____ fastening _____

    Shingles _____ grade _____ type _____ size _____ exposure _____ fastening _____

    Stucco _____ thickness _____ Lath _____ weight _____

    Masonry veneer _____ Sills _____ Lintels _____ Base flashing _____

Masonry: ☐ solid ☐ faced ☐ stuccoed; total wall thickness _____ facing thickness _____ facing material _____

                    Backup material _____ thickness _____ bonding _____

    Door sills _____ Window sills _____ Lintels _____ Base flashing _____

    Interior surfaces: dampproofing _____ costs of _____ furring _____

Additional information: _____

Exterior painting: material _____ number of coats _____

Gable wall construction: ☐ same as main walls   ☐ other construction _____

6) FLOOR FRAMING:

    Joists: wood, grade, and species _____ other _____ bridging _____ anchors _____

    Concrete slab: ☐ basement floor   ☐ first floor   ☐ ground supported ☐self-supporting: mix _____ thickness ___

    reinforcing _____ insulation _____ membrane _____

    Fill under slab: material _____ thickness _____ Additional information: _____

7) FINISH FLOORING: (Wood only. Describe other finish flooring under item 21.)

LOCATION	ROOMS	GRADE	SPECIES	THICKNESS	WIDTH	BLDG. PAPER	FINISH
First Floor							
Second Floor							
Attic Floor							

8) PARTITION FRAMING:

    Studs: wood, grade, and species_____ size and spacing _____ Other _____

    Additional information: _____

9) CEILING FRAMING:

    Joists: wood, grade, and species _____ Other _____ Bridging _____

    Additional information: _____

10) ROOF FRAMING:

    Rafters: wood, grade, and species _____ Roof trusses: grade and species _____

    Additional information: _____

11) ROOFING:

Sheathing: wood, grade, and species _____ ☐ solid ☐ spaced

Roofing _____ grade _____ size _____ type _____

Underlay _____ weight or thickness _____ size _____ fastening _____

Built-up roofing _____ number of plies _____ surfacing material _____

Flashing: material _____ gage or weight _____ ☐ gravel stops ☐ snow guards

Additional information: _____

12) DECORATING: (Paint, wallpaper, etc.)

ROOMS	WALL FINISH MATERIAL & APPLICATION	CEILING FINISH MATERIAL AND APPLICATION
Kitchen		
Bath		
Other		

13) INTERIOR DOORS AND TRIM:

Doors: type _____ material _____ thickness _____

Door trim: type _____ material _____ Base: type _____ material _____ thickness _____

Finish: doors _____ trim _____

Other trim (item, type, and location) _____

Additional information: _____

14) WINDOWS:

Windows: type _____ make _____ material _____ thickness _____

Glass: grade _____ ☐ sash weights ☐ balances, type _____ head flashing _____

Trim: type _____ material _____ paint _____

Weatherstripping: type _____ material _____ storm sash, number _____

Screens: ☐ full ☐ half, type _____ number _____ screen cloth material _____

Basement windows: type _____ material _____ screens, number _____ storm sash, number _____

Special windows _____

Additional information: _____

15) CABINETS AND INTERIOR DETAIL:

Kitchen cabinets, wall units: material _____ lineal feet of shelves _____ shelf width _____

Base units: material _____ counter top _____ edging _____

Back and end splash _____ Finish of cabinets _____

Medicine cabinets: make _____ model _____

Other cabinets and built-in furniture _____

Additional information: _____

16) STAIRS:

STAIR	TREADS		RISERS		STRINGERS		HANDRAIL		BALLISTERS	
	Mat.	Thickness	Mat.	Thickness	Mat.	Size	Mat.	Size	Mat.	Size
Basement										
Main										
Attic										

Additional information:

17) PLUMBING:

FIXTURE	NUMBER	LOCATION	MAKE	MFG'S FIXTURE IDENTIFICATION NO.	SIZE	COLOR
Sink						
Lavatory						
Water Closet						
Bathtub						
Shower over Tub						
Stall Shower						

18) HEATING:

☐ Hot water ☐ Steam ☐ Vapor ☐ One-pipe system ☐ Two-pipe system

☐ Radiators ☐ Convectors ☐ Baseboard radiation   Make and model _____

Radiant panel: ☐ floor ☐ wall ☐ ceiling   Panel coil: material _____

☐ Circulator ☐ Return pump   Make and model _____

Boiler: Make and model _____

Additional information _____

Warm air: ☐ Gravity ☐ Forced   Type of system _____

Duct material: supply _____ return _____ insulation _____ thickness _____

Furnace: make and model _____ input _____ Btuh; output _____ Btuh

Additional information _____

☐ Space heater ☐ floor furnace ☐ wall heater   input _____ Btuh; output _____ Btuh; number units _____

Make and model _____ Additional information _____

Controls: Makes and types

Additional information: _____

Fuel: ☐ coal ☐ oil ☐ gas ☐ liq. pet. gas ☐ electric ☐ other _____ storage capacity _____

Additional information _____

Firing equipment furnished separately: ☐Gas burner, conversion type ☐ Stoker: ☐hopper feed ☐bin feed

Oil burner: ☐ pressure atomizing ☐ vaporizing _____

Make and model _____ Control _____

Additional information _____

Electric heating system: type _____ input _____ watts @ ____ volts; output _____ Btuh

Additional information _____

Ventilating equipment: attic fan, make and model _____ capacity _____ cfm

kitchen exhaust fan, make and model _____

Other heating, ventilating, or cooling equipment _____

19) ELECTRIC WIRING:

Service: ☐ overhead ☐ underground   Panel:☐ fuse box ☐ circuit-breaker, make _____ AMPs _____ No. circuits ____

Wiring: ☐ conduit ☐armored cable ☐ nonmetallic cable ☐ knob and tube ☐ other _____

Special outlets: ☐ range ☐ water heater ☐ other _____

☐ Doorbell ☐ Chimes; Push-button locations _____

Additional information: _____

20) LIGHTING FIXTURES:

Total number of fixtures _____ Total allowance for fixtures, typical installations, $ _____

Additional information: _____

21) INSULATION:

LOCATION	THICKNESS	MATERIAL, TYPE, AND METHOD OF INSTALLATION	VAPOR BARRIER
Roof			
Ceiling			
Wall			
Floor			

HARDWARE: (make, material, and finish) _____

_____

_____

_____

22) MISCELLANEOUS: _____

_____

_____

_____

_____

_____

PORCHES: _____

GARAGES: _____

23) WALKS AND DRIVEWAYS:

Driveway: width _____ base material _____ thickness _____ surfacing material _____ thickness _____

Front walk: width _____ material _____ thickness _____

Service walk: width _____ material _____ thickness _____

Steps: material _____ treads _____ "   risers _____ "   Check walls _____

OTHER ONSITE IMPROVEMENTS: (Include items such as unusual grading, drainage structures, retaining walls, fence, railings, and accessory structures)

_____

_____

_____

_____

Date _____   Signature _____

Signature _____

QUESTIONS FOR CHAPTER 8

Q-1   What items are included in the construction specifications?

A-1   _____

_____

_____

Q-2   Why should the contractor be familiar with the specifications?

A-2   _____

_____

_____

_____

Q-3   What does the term same or equal quality mean?

A-3   _____

_____

_____

Q-4   In construction specifications, what type of information is included in the general information section?

A-4   _____

_____

_____

Q-5   What information is included in the carpentry section?

A-5   _____

      _____

      _____

Q-6   Name three components included in the plaster and lath section.

A-6   _____

      _____

      _____

Q-7   What is included in the electrical category?

A-7   _____

      _____

      _____

Q-8   Why are guarantees included in the construction specifications?

A-8   _____

      _____

      _____

CHAPTER 9
SCHEDULES

Schedules provide detailed information about particular items on blueprints. This information is presented in the form of a table, with reference numbers corresponding to numbers on the blueprint. When the number on the blueprint is referenced to the schedule, more information concerning that particular item may be found on the schedule.

SCHEDULES

Schedules are necessary in addition to the construction specifications, because not all the data found in schedules pertains to the construction specifications. This information includes the specific styles of components (i.e., doors and windows), and if all this information were included on the construction specifications, they would be very long. Schedules are also necessary, because the construction specifications are not usually found on the work site, so schedules are used as an abbreviated form of these specifications.

Each item in a schedule has a symbol or reference number, in order to show its location on the blueprint. Most schedules include the quantity required for each item, as well as the size, color, and any other pertinent information concerning that item. Schedules are part of the entire legal contract, so all parties involved in the construction must grant permission in writing if a change is to be made.

A contractor bids for a project (see Chapter 1) based on the blueprint and the schedule. This allows each contractor to know exactly what types of materials will be used for the project, so the bidding is fair. For example, if two HVAC contractors are bidding for the same project, each will make a bid based on the estimated cost of materials and labor outlined in the blueprints and schedules. For this reason, it is necessary for the contractor to pay very close attention to the blueprint and the schedule.

Almost every item found in a building has a schedule associated with it. For example, there are lighting fixture schedules, door and window schedules, exhaust fan schedules, etc. Normally, the architect and engineers create each schedule, in order to avoid any misunderstandings that may arise with the contractor. There is a schedule that combines most of the other schedules and presents all the information in one document. This is called the material schedule.

Material Schedule. The material schedule (also called a material list or bill of materials) is a list of the approximate quantities of all the materials that are to be used in the construction. This schedule is just an approximation, and it should only be used to estimate the quantity needed. To find the actual quantity needed it is necessary to consult the more specific schedule. As there are so many schedules involved in a particular project, it is impossible to compile a totally accurate material schedule. Figure 9-1 is an example of a material schedule.

LINE NO.	ITEM COLUMN	QUANTITY AND MEASUREMENT	MATERIAL: TYPE AND/OR SIZE	UNIT COST	TOTAL COST
1	MASONRY				
2	85 Lin. Ft. 90 x 15	324 Sacks	Cement		
3	150 Lin. Ft. 9 x 47	162 Pos.	4" Drain Tile		
4	1 Pad 3' mix 7' mix	8 Pos.	4" Vitrified Crock		
5	Masonry Block Walls				
6		80 Pos.	12 x 8 x 16" Grade Blocks		
7		9 Pos.	12 x 8 x 16" Corner Blocks		
8		650 Pos.	12 x 8 x 16" Regular Blocks		
9		45 Pos.	8 x 8 x 16" Solid Blocks		
10		150 Pos.	8 x 8 x 16" Solid "L" Blocks		
11		14 Cu. Yds.	50/50 Mason Sand		
12		60 Sacks	Mortar		
13		20 Sacks	Cement		
14		23 Gals.	Asphalt Foundation Costing		
15		1 Pos.	6" Diam. Steel Furnace Thimble		
16		8 Pos.	12 x 12 Flue Lining		
17	Fireplace & Veneer				
18		3 Pos.	52" Dome Damper		
19		275 Pos.	Firebrick		
20		1 Unit	40 x 95" Double Barbecue		
21		8 Gals.	Asphalt Foundation Coating		
22		2700 Sq. Ft.	4" Cut Stone Veneer		
23	CONCRETE SLABS				
24		4000 Sq. Ft.	8 x 8 #6/6 Welded Wire Mesh		
25		250 Sacks	Cement		

Figure 9-1. Material Schedule

ROOMS	FLOOR			CEILING			WALL			BASE			REMARKS
	SLATE	CORK TILE	CARPET	PLASTER	WOOD PANEL	DIATO	WALL PAPER	PLASTER	WOOD PANEL	ASPHALT	RUBBER	WOOD	
BEDROOM 1			X	X			X					X	
BEDROOM 2			X	X		X			X			X	
BATH 1	X				X			X		X			
KITCHEN		X			X		X		X	X			
DINING			X			X	X				X		
LIVING			X		X				X			X	
HALL	X				X		X			X			

Figure 9-2. Material Schedule as a Checklist

Material schedules can also be used as a checklist, to ensure all the finishing touches in a particular building match. This version of a material schedule summarizes some of the many other schedules available into one table. The contractor only has to look at this one table to make sure all the items match properly. Figure 9-2 is an example of this material schedule.

Other Schedules. As previously stated, almost every item in a building uses a schedule. For example, window and door schedules provide information

concerning their heights, widths, types, screens, etc., as all this information would be impossible to place on a blueprint.

HVAC schedules encompass a variety of elements. Figure 9-3 is an example of a fan schedule, which the HVAC contractor would need, and Figure 9-4 shows the air terminal schedule.

While most of the terminology found on a schedule is self-explanatory, there are some terms that may not be familiar. For example, Figure 9-5 shows a radiation schedule, which is used to describe hot-water or steam baseboard units. The terms used in this figure need some explanation.

FAN SCHEDULE								
SYMBOL	AREA SERVED	CFM	S.P. IN.W.G.	MOTOR			MFG.	CAT. NO.
				HP	RPM	VOLT.		
KEF-1	HOOD EXHAUST	4375	2.5	3	1750	460/3/60	GREENHECK	CUBE HP-24-30
KEF-2	HOOD EXHAUST	1625	2	2	1390	460/3/60		CUBE HP-18-15
EF-5	MICROFILM ROOM	1000	.375	1/4	1750	120/1/60		CUBE 14-4
SF-1	BOILER ROOM	400	.5	1/2	1050			CSP-60
SF-2								
SF-3								
RAF-1	CAFETERIA	4125	1.0	2	1750			BSQ 21-15

Figure 9-3. Fan Schedule

AIR TERMINAL				
NECK SIZE	CFM RANGE	TOTAL PRESSURE IN.W.G.	FACE SIZE	BASIS OF DESIGN "TITUS PRODUCTS"
6"	50 - 120	.05 MAX	24 x 24"	MODEL PAS
8"	150 - 250	.10 MAX	24 x 24"	MODEL PAS
10"	260 - 380	.10 MAX	24 x24"	MODEL PAS
12"	390 - 600	.15 MAX	24 x 24"	MODEL PAS
14"	665	.12 MAX	24 x 24"	MODEL PAS
22 x 22"	280 - 1200	NEG. SP - .05	24 x 24"	MODEL PAR

Figure 9-4. Air Terminal Schedule

RADIATION SCHEDULE									
SYMBOL	SERVING	CAPACITY MBH	ENCLOSURE LENGTH	ACTIVE LENGTH	RATING - BTU LIN.FT.	PROTOTYPE	TYPE OF ENCLOSURE BACK	TYPE OF SLEEVE	REMARKS
	LOBBY	21.9	28'	23'	800	ARCH SILL MNS-3	UNFINISHED	INTERNAL SLEEVE AT MULLIONS	
	OFFICE 1	9.8	17'	16'	780	ARCH SILL MNS-4	FINISHED	INTERNAL SLEEVE AT MULLIONS	
	OFFICE 2	26.6	31'	28'	820	ARCH SILL MNS-5	FINISHED	NONE	ENCLOSURE FULL LENGTH OF WALL
	LUNCH ROOM	57	15'	10'	900	ARCH SILL RTL-4	UNFINISHED	INTERNAL SLEEVE AT MULLIONS	
	OFFICE 3	28	38-1/2'	36'	750	ARCH SILL RTL-3	FINISHED	NONE	
	BATHROOM	19.3	4'	3'	745	ARCH SILL RTL-2	UNFINISHED	INTERNAL SLEEVE AT MULLIONS	

Figure 9-5. Radiation Schedule

• The *symbol* column simply indicates where on the blueprint this unit is found.

• The *serving* column states which room or area receives heat from the unit.

• In the *capacity MBH* column, the MBH stands for one thousand Btu per hour. Basically then, this column specifies each unit's heat output.

• The *enclosure length* shows the actual length of each radiator.

• The *active length* column defines the length of radiation piping inside the baseboard unit. The difference between the enclosure length and the active length is the room necessary for pipe connections.

• The *prototype* column shows the manufacturer's description of the unit to be used.

• The *remarks* column is there for the engineer to add any further notes that may help clarify the specifications for the unit.

QUESTIONS FOR CHAPTER 9

Q-1   What is a schedule used for?

A-1   _____

_____

_____

Q-2   Why are schedules necessary?

A-2   _____

_____

_____

_____

Q-3   What is the purpose of the reference numbers listed on the schedule?

A-3   _____

_____

_____

Q-4   Who compiles the schedules?

A-4   _____

Q-5   What is a material schedule?

A-5   _____

       _____

       _____

Q-6   How can a material schedule be used as a checklist?

A-6   _____

       _____

       _____

Q-7   What does MBH stand for?

A-7   _____

ASSIGNMENT 1
HOUSE PLANS

The first assignment illustrates some of the information a contractor must know before proceeding with the construction of a house.

HOUSE BLUEPRINTS

There are several different types of blueprints used in the set of house plans for this assignment. It is necessary to study all the blueprints included in order to answer the questions that follow.

The main floor plan, Figure A-1, shows the location of each projection-style air supply and return grille. This plan also shows the location of the outdoor air

conditioning condenser. Figure A-2 shows the same house outline as Figure A-1, but Figure A-2 shows the duct system and the location of the furnace and air conditioning evaporator. A third drawing, Figure A-3, shows the location of the electrical components related to the HVAC system only.

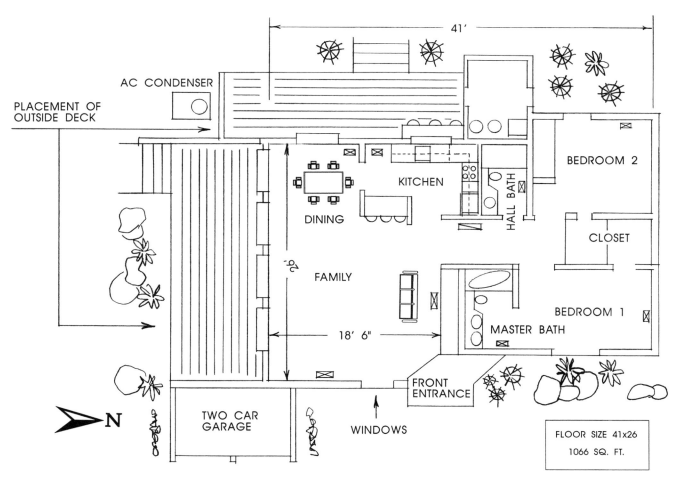

Figure A-1. Main Floor Plan

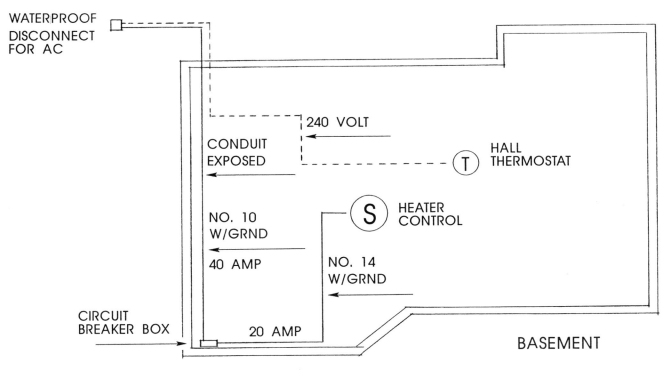

Figure A-2. Duct System and Heating-Cooling System Location

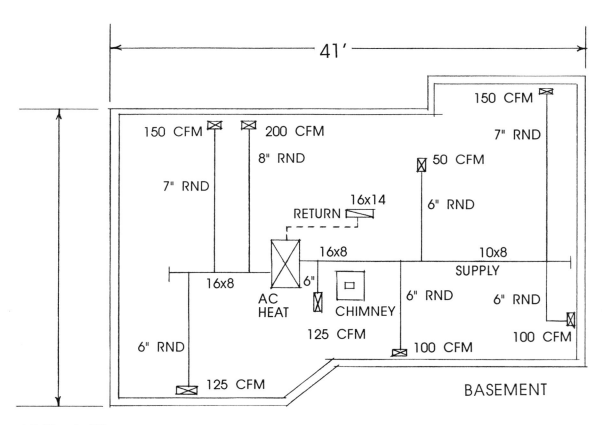

Figure A-3. Electrical Plan

Along with the blueprints already shown, the HVAC
contractor also receives a schedule, which is shown in
Figure A-4.

HVAC MATERIAL SCHEDULE			
MANUFACTURER	**MODEL**	**BTU**	**VOLT**
RUUD FURNACE	UGEB10EEGS	100,000	120
RUUD CONDENSER	UAFD-030JA	30,000	240
RUUD EVAPORATOR	UCEH-Z031	30,000	
HART & COOLEY	**REGISTERS**	**SIZE**	**CFM**
	SUPPLY A611	8" X 4"	50
		10" X 6"	100
		10" X 8"	125
		12" X 8"	150
		14" X 6"	200
	RETURN 673 WITH FILTER	16" X 14"	1000

Figure A-4. HVAC Schedule

QUESTIONS FOR ASSIGNMENT 1

Q-1   For each of the rooms listed below, supply the following information.

	Kitchen:	Dining room:	Family room:	Hall bath:	Master bath:	Bedroom 1:	Bedroom 2:
Number of supply registers.							
Size of each.							
Model number for each.							
CFM to be delivered by each.							
Total cfm for each room.							

Q-2.   Where is the return register located?

A-2   _____

_____

_____

Q-3.   What type and size is the return register?

A-3   _____

_____

_____

Q-4. How many different kinds of round ducts are used and what are their sizes?

A-4 _____

_____

_____

Q-5. List each register size and the size of the round duct that supplies it.

A-5 _____

_____

_____

Q-6. Where is the air conditioning condenser located?

A-6 _____

_____

_____

Q-7. List the model and Btu size for the condenser.

A-7 _____

_____

_____

Q-8. List the model and Btu size for the furnace.

A-8 _____

_____

_____

Q-9. List the model and Btu size for the evaporator.

A-9 _____

_____

_____

Q-10. List the circuit breaker and wire size for the condenser.

A-10 _____

_____

_____

Q-11. List the circuit breaker and wire size for the furnace.

A-11 _____

_____

_____

OFFICE BUILDING PLANS

This assignment illustrates some of the information necessary when constructing an office building.

OFFICE BUILDING BLUEPRINTS

Much like following the house blueprints in the previous assignment, there are several different types of office building blueprints that must be studied in order to answer the questions that follow.

Figure B-1 illustrates the main floor of an office building. Included in this blueprint is the location of the ceiling registers for air conditioning and the baseboard units for heating purposes. Figure B-2 shows the duct system in the attic, including the supply and return registers. Also pictured is the approximate location of the air conditioner. Figure B-3 shows the hot water piping system located in the basement of the building. The hot water supply (HWS) lines and the hot water return (HWR) lines are shown, as is the location of the hot water boiler. Figure B-4 shows the HVAC schedule for the office building project.

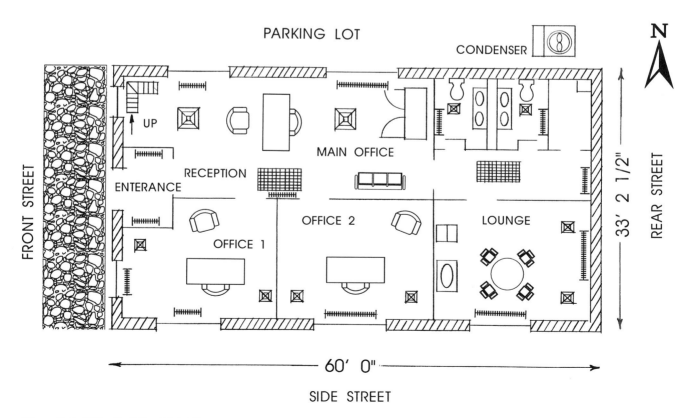

Figure B-1. Office Building Main Floor

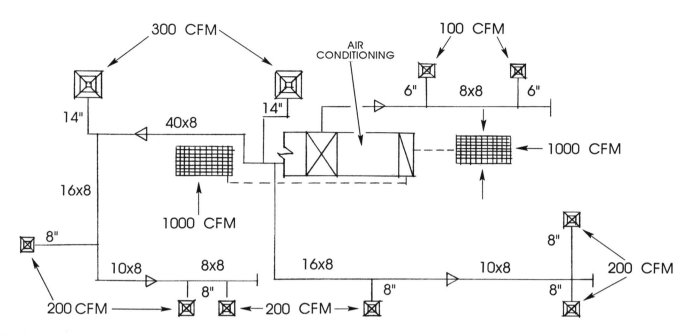

Figure B-2. Attic Duct System for Air Conditioning

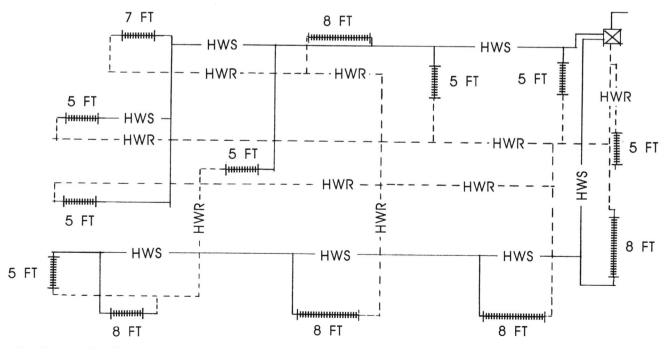

Figure B-3. Basement Hot Water Piping System for Heating

HVAC MATERIAL SCHEDULE			
MANUFACTURER	**MODEL**	**BTU**	**VOLT**
VAILLANT BOILER	GA92-100EI	100,000	120
BRYANT MFG. CONDENSER	565BB060	60,000	240
BRYANT MFG. BLOWER COIL	517A090	60,000	
HART & COOLEY	**CEILING DIFFUSERS**	**SIZE**	**CFM**
	AMD - 3WAY	8" x 8"	100
		10" x 10"	200
		12" x 12"	300
	RETURN RHF45FILTER	24" x 48"	1000
HEATRIM AMERICAN HYDRONIC BASEBOARD		**PRE-ASSEMBLED SIZE**	**BTU**
		5 feet	3000
		7 feet	4200
		8 feet	4800

Figure B-4. HVAC Schedule

QUESTIONS FOR ASSIGNMENT 2

Q-1   How many areas in this building require cooling?

A-1   _____

Q-2   List the areas requiring cooling.

A-2   _____

_____

_____

Q-3   How many areas in the building require heating?

A-3   _____

Q-4   List the areas requiring heating.

A-4   _____

_____

_____

Q-5   List the number and size of registers for each of the following areas:

	Number of registers	Size of registers
Reception Area		
Bathrooms		
Main Office		
Office 1		
Office 2		
Lounge		

Q-6   List the number, size, and Btu capacity of the baseboard units for each of the following areas:

	Number of radiators	Size of radiators	Btu capacity
Reception Area			
Bathrooms			
Main Office			
Office 1			
Office 2			
Lounge			
Entrance			
Rear Hall			

Q-7   How many return registers are located on the main floor?

A-7   _____

Q-8   What is the model number and size of the return register(s)?

A-8   _____

_____

_____

Q-9   What is the total cfm for the building?

A-9   _____

Q-10 List the duct sizes being used.

A-10 _____

_____

_____

Q-11 Where is the condenser located?

A-11 _____

_____

_____

Q-12 Where is the central air conditioning unit containing the evaporator (also called an air handler) located?

A-12 _____

_____

_____

Q-13 Where is the boiler located and what is its size in Btu, model number and manufacturer?

A-13 _____

_____

_____

Q-14 List the manufacturer for the baseboard units.

A-14 _____

_____

_____

Q-15 Who manufactures the condenser?

A-15 _____

BOILER PIPING FOR BUILDING HEATING

HVAC contractors must understand the piping associated with hydronic systems. This assignment focuses on the layout of this piping.

MECHANICAL ROOM BLUEPRINT

Figure C-1 shows the blueprint concerning the piping for a hydronic system. Figure C-2 shows a sectional view of the boilers. The scale for both drawings is 1/4" = 1'-0".

These blueprints use many plumbing and piping symbols to represent actual parts. Also, the letters HWS and HWR are used to indicate the hot water supply line and the hot water return line.

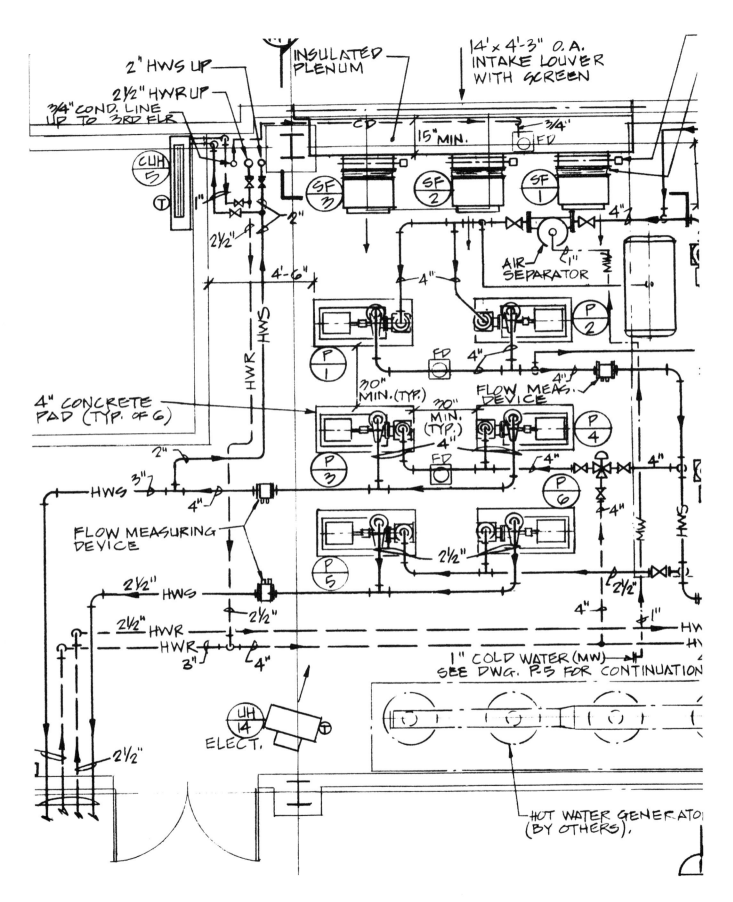

Figure C-1. Hydronic System Piping

MECHANICAL

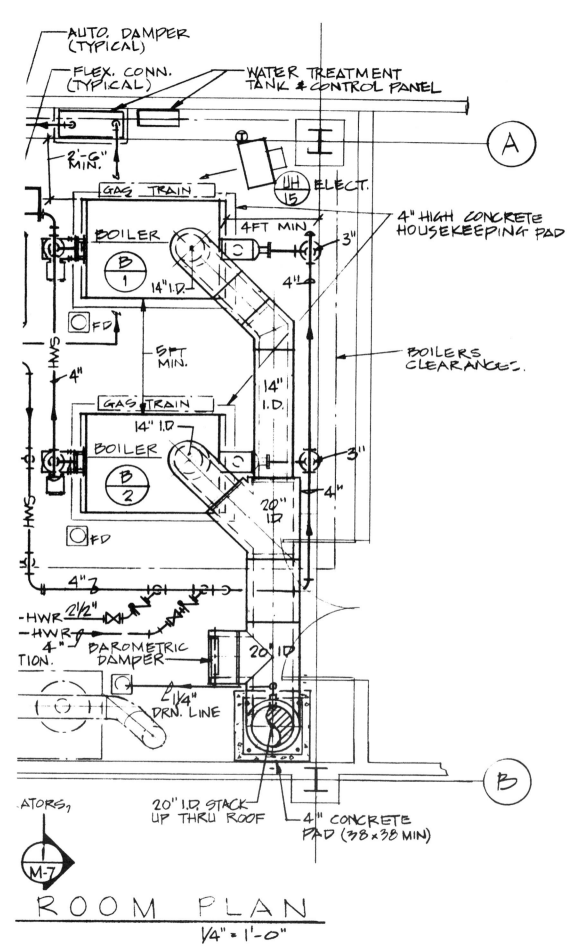

AUTO. DAMPER
(TYPICAL)

FLEX. CONN.
(TYPICAL)

WATER TREATMENT
TANK & CONTROL PANEL

Ⓐ

2'-6"
MIN.

GAS TRAIN

UH
15 ELECT.

4" HIGH CONCRETE
HOUSEKEEPING PAD

BOILER

B
1

14"I.D.

4FT MIN

3"

4"

BOILERS
CLEARANCES.

FD

HWS

4"

5FT
MIN.

GAS TRAIN

14" 1.D

BOILER

B
2

14" 1.D.

20"
1.D

3"

4"

HWS

4" 2

HWR 2½"

HWR

4"

TION.

BAROMETRIC
DAMPER

20" 1.D

1¼"
DRN. LINE

20"1.D

Ⓑ

ATORS,

M-7

20" I.D. STACK
UP THRU ROOF

4" CONCRETE
PAD (38×38 MIN)

ROOM   PLAN
¼" = 1'-0"

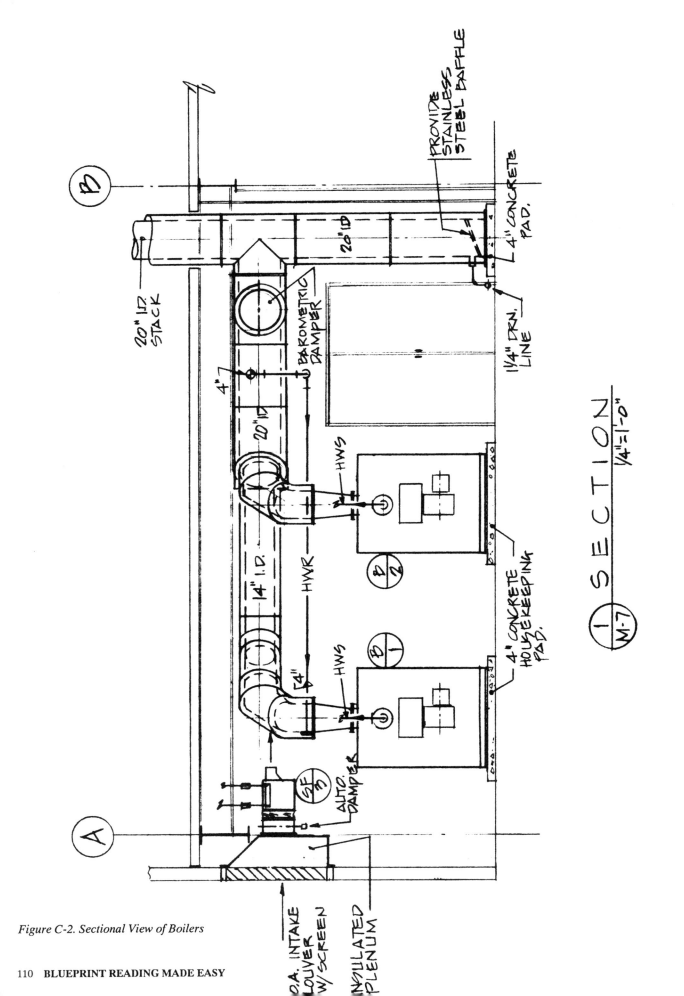

Figure C-2. Sectional View of Boilers

QUESTIONS FOR ASSIGNMENT 3

Q-1 What is the pipe size used for the supply lines?

A-1 _____

_____

_____

Q-2 What is the pipe size used for the return lines?

A-2 _____

_____

_____

Q-3 The following symbols are used in the drawing, name them:

a. ⊲⊳ _____

b. ▷ _____

c. ⌐ _____

d. ⟶○ _____

e. ⊥ _____

Q-4 Excluding the concrete pads, how tall are the boilers?

A-4 _____

_____

_____

Q-5 How does fresh air enter the mechanical room?

A-5 _____

_____

_____

Q-6 How wide is each boiler?

A-6 _____

_____

_____

Q-7 Will the boilers fit through the mechanical room door? How wide is the door?

A-7 _____

_____

_____

Q-8 What size flue pipe is needed?

A-8 _____

_____

_____

Q-9 How far apart are the boilers, excluding the concrete pads?

A-9 _____

_____

_____

ASSIGNMENT 4
KITCHEN PLANS

The kitchen plan is but a small part of a total building plan. The blueprints in this assignment cover some of the information required when constructing a kitchen.

KITCHEN BLUEPRINTS

Figure D-1 shows the floor plan of part of a building's first floor. The scale for this drawing is 1/16" = 1'-0". This is a very small scale, so the blueprint looks very crowded.

In Figure D-1, there is an area labeled M-6 on the right of the drawing. This area is the kitchen duct system, which is shown in greater detail in Figure D-2. This drawing uses the larger scale of 1/4" = 1'-0".

Figures D-3, D-4, and D-5 are taken directly from Figure D-2, but it is easier to answer the questions below with these sections separated. These drawings are also to the scale of 1/4" = 1'-0".

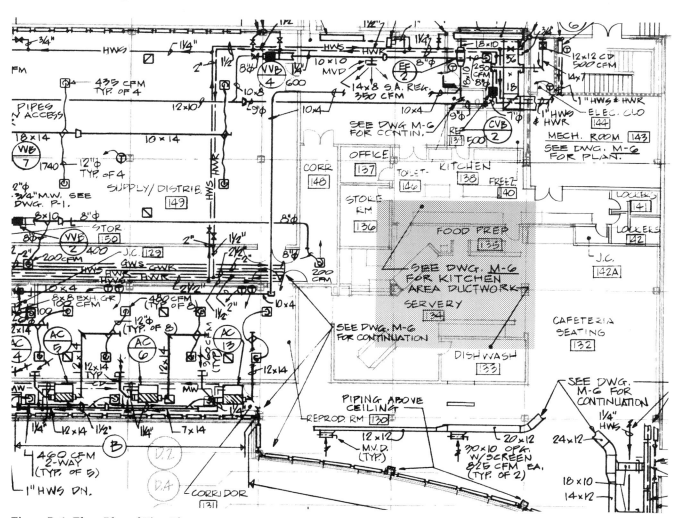

Figure D-1. Floor Plan of First Floor

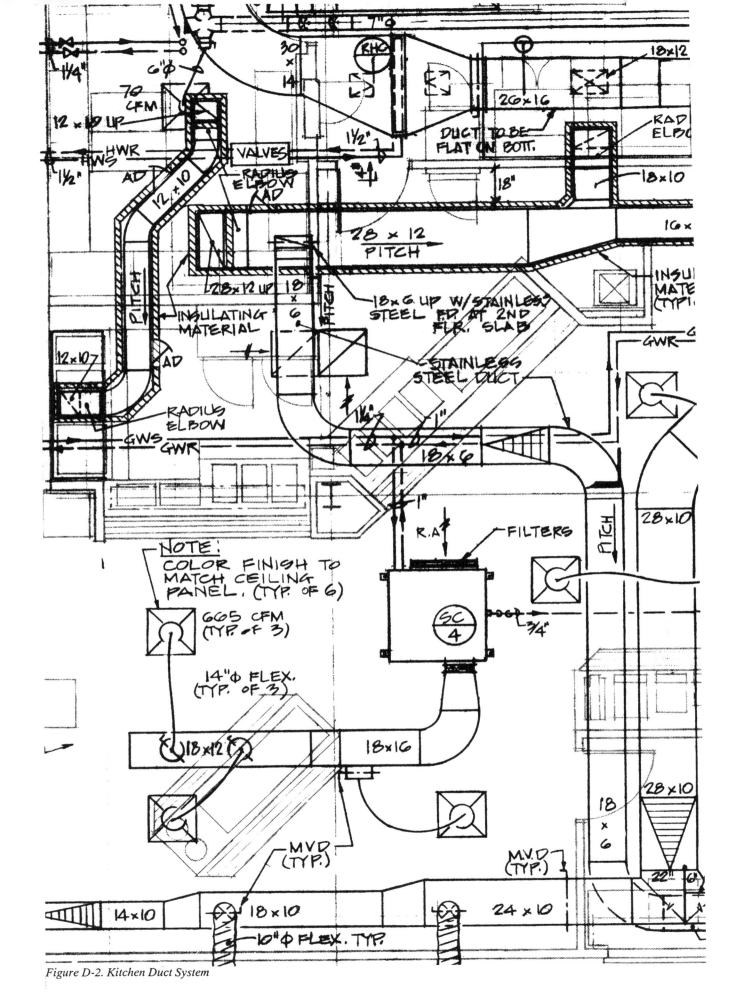

Figure D-2. Kitchen Duct System

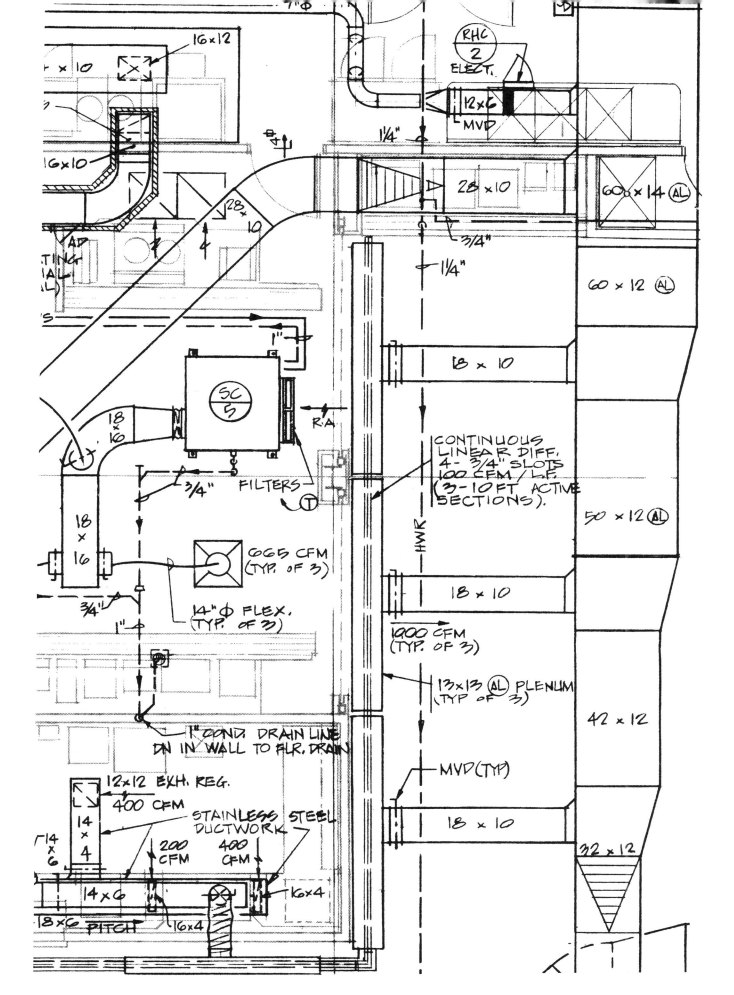

16×12

RHC
2
ELECT.

12×6
MVD

28×10

60×14 (AL)

60×12 (AL)

1¼"

3/4"

1¼"

18×10

SC
5

R.A.

CONTINUOUS
LINEAR DIFF.
4-¾" SLOTS
100 CFM/L.F.
(3-10 FT ACTIVE
SECTIONS).

50×12 (AL)

18×16

3/4"

FILTERS

T

665 CFM
(TYP. OF 3)

HWR

18×10

18
×
16

1000 CFM
(TYP. OF 3)

3/4"

14"Φ FLEX.
(TYP. OF 3)

13×13 (AL) PLENUM
(TYP OF 3)

42×12

1" COND. DRAIN LINE
DN IN WALL TO FLR. DRAIN

MVD (TYP)

12×12 EXH. REG.

400 CFM

14
×
4

STAINLESS STEEL
DUCTWORK

18×10

14
×
6

200
CFM

400
CFM

32×12

14×6

16×4

18×6   PITCH

16×4

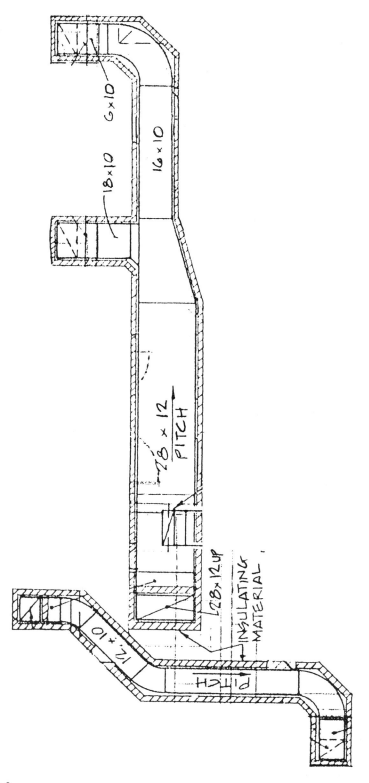

Figure D-3. Section of D-2

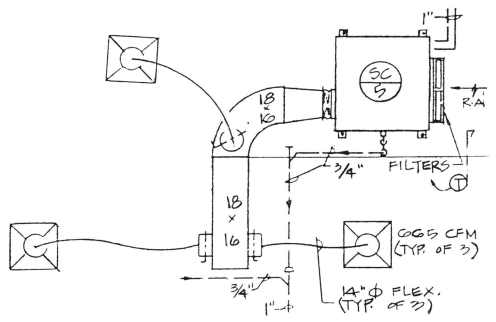

Figure D-4. Section of D-2

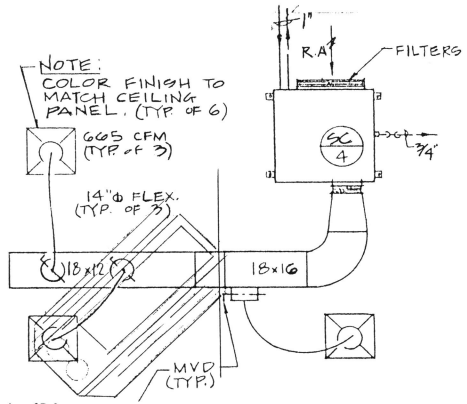

NOTE:
COLOR FINISH TO
MATCH CEILING
PANEL. (TYP. OF 6)

665 CFM
(TYP. OF 3)

14"Φ FLEX.
(TYP. OF 3)

18 x 12

18 x 16

R.A.

FILTERS

SC
4

1"

3/4"

MVD
(TYP.)

Figure D-5. Section of D-2

QUESTIONS FOR ASSIGNMENT 4

Q-1   In Figures D-4 and D-5, two handlers are labeled SC-5, and SC-4, respectively. What diameter condensate line is listed for these units?

A-1   _____

_____

_____

Q-2   What is the size of the condensate line when it empties into the drain?

A-2   _____

_____

_____

Q-3   How many ceiling registers are supplied for each unit and how many cfm does each register supply?

A-3   _____

_____

_____

Q-4   What type of ductwork is used to connect the ceiling registers to the supply duct? What is its size?

A-4   _____

_____

_____

Q-5   Is there a return duct system connected to these systems? If yes, what is the size?

A-5   _____

      _____

      _____

ANSWERS

CHAPTER 1

A-1   Construction prints are called blueprints, because the original prints were white lines on blue paper.

A-2   Whiteprinting, photocopying, microfilming, or using a plotter are the most common reproducing methods.

A-3   Microfilm is used, because it preserves the blueprint, and its small size requires less space for storage.

A-4   Damaged blueprints can be restored by using a filter on a photocopying camera to compensate for a stained original blueprint. Or, blueprints can be cleaned using a water-alcohol solution.

A-5   Working blueprints is the name of the complete set of blueprints.

A-6   It is necessary to understand what is drawn on the blueprint, because misunderstood details can lead to construction delays, additional costs or a legal battle.

A-7   No, each subcontractor will not receive a complete set of construction prints.

A-8   CAD stands for computer-aided design, and a person can use this program to design a structure on a computer.

A-9   An architect creates the actual blueprint.

A-10   A plotter or printer transfers the drawings from computer to paper.

A-11   Yes, a contractor may make changes to the blueprint, but not without the approval of the architect.

A-12   In order to make changes, the blueprint should be submitted to the architect and engineers, and if necessary, the building inspector, for approval.

CHAPTER 2

A-1   Symbols simplify the work of an architect or an engineer.

A-2   EARTH   ROCK

A-3   THERMOSTAT   MOTOR CONNECTION

A-4   Smoke detector, SD; Exit light; X

A-5   EXHAUST FAN   HEAT REGISTER   `R`

A-6   SUPPLY AIR DUCT   HEAT PUMP   `HP`

A-7   In a plumbing blueprint, symbols represent piping, fixtures, valves, and fittings.

A-8   COUPLING   —┼—   EXPANSION JOINT

A-9   Elevation views and floor plans use symbols most often.

A-10   STEEL TUBE

A-11   It is necessary to know construction synonyms, because different people use different terms to describe the same object.

A-12   The common abbreviation for British Thermal Unit is Btu.

A-13   Faucet - Tap; Wall - Partition

CHAPTER 3

A-1   A section line separates different materials in a sectional drawing.

A-2   An object line defines the boundaries of the building, including exterior and interior walls, patios, decks, etc.

A-3   DIMENSION LINE

A-4   The center line identifies the center of symmetrical objects and the exact location of a building component can be determined by measuring from the center line.

A-5   CENTER LINE

A-6   A break line may be used when a particular component remains the same for a long distance on a blueprint.

A-7   The line legend is necessary, so anyone reading the blueprint knows what the lines represent.

A-8   A phantom line can represent a building addition, moveable walls, or fixture positions.

A-9   Stair lines indicate in which direction the stairs are heading.

A-10   Lines are necessary to relate important information concerning all aspects of a particular building to the contractor, as well as to anyone else involved with the building construction.

CHAPTER 4

A-1   Scales are necessary, because it is not feasible to draw a building at its full size.

A-2   An architect's scale uses a certain ratio in order to create an actual representation of the building in a reduced size.

A-3   On a triangular scale there are a maximum of 6 different scales available.

A-4   Not-to-scale drawings usually show unusual features or installation procedures.

A-5   The building will measure 6-1/4 inches (1/8 x 50 = 6-1/4).

A-6   To create clear drawings, architects draw the dimension lines away from the building lines; they make sure all blueprints match; they show wall thickness, and they mark all masonry openings.

A-7   An engineer's scale differs from the architect's scale in that each inch in an engineer's scale is divided into 10, 20, 30, 40, 50, or 60 equal units.

A-8   The actual measurement is 240 feet (2÷1/4 = 8; 8 x 30 = 240).

A-9   A meter measures distance, a kilogram measures weight, and a liter measures volume.

A-10   A metric scale is based on increments of 10 and the architect's scale is based on increments of 12.

A-11   Dual dimensions mean that both the metric scale and the standard U.S. dimensions are shown.

A-12   An open-divided scale has the main units numbered along the entire length of the scale with a fully-subdivided extra unit located at one end.

A-13   A fully-divided scale fully subdivides each main unit along the entire length of the scale.

CHAPTER 5

A-1   The different styles of construction drawings are pictorial, orthographic, and diagrams or schematics.

A-2   Perspective drawings are the most realistic.

A-3   Isometric drawings are preferred.

A-4   The orthographic drawing shows the actual arrangement and views of particular objects.

A-5   A mechanical plan requires drain and water supply piping diagrams, air duct and grille diagrams, elevation diagrams of the drainage system, schematic wiring diagrams of the control circuits and diagrams illustrating pneumatic controls.

A-6   A diagram is a drawing used to simplify components and their related connections.

A-7   A schematic diagram is an actual representation of the electronic controls for the electronic systems in a building.

A-8   No, diagrams are not drawn to scale.

A-9   An oblique drawing is a type of pictorial drawing, drawn to give a somewhat exact representation of the object being drawn.

A-10   An isometric drawing is a type of pictorial drawing, and shows two sides and a top or bottom of a building or object.

CHAPTER 6

A-1   Location of building, property boundaries, trees and other permanent objects are usually found on a site plan.

A-2   A land surveyor, initially, and sometimes an architect draws the site plan.

A-3   Footings are made of concrete and distribute the weight of the building under the ground line. Sills provide a base, so the exterior walls can attach to the foundation.

A-4   The floor plan gives the most information.

A-5   A working-drawing floor plan provides the contractor with the information necessary to correctly interpret the architect's design.

A-6   A sectional drawing details the internal construction of a particular building, or part of the building.

A-7   A cutting plane is what the architect imagines is cutting a building into sections.

A-8   Dotted lines are not acceptable in a sectional drawing because they are not clear on paper, and they can be confusing.

A-9   Dead weight is the weight of the materials used to construct the building. It is important, because the architect bases part of the framing plan around dead weight.

A-10   Mechanical plans include information concerning the installation of heating, ventilating, air conditioning and plumbing systems.

CHAPTER 7

A-1   An engineer's scale is normally used for site plans.

A-2   Dimension lines are placed either outside the property line or actually on the property line.

A-3   Slab foundation, T-foundation, and column foundation are the three foundation types.

A-4   Complete basement plans show laundry and bathroom fixtures, floor drains, fireplaces, and the locations of posts and girders.

A-5   When a building component is too small for dimension numerals, the dimension is positioned outside the extension lines. Or, the arrowheads may be placed outside the extension lines.

A-6   Wall thickness does not affect room dimensions.

A-7   The datum line is a constant horizontal plane. In elevation drawings, it is often considered to be at sea level.

A-8   Rise is the vertical measurement of the roof and run is the horizontal measurement of the roof. Together they determine the pitch, or angle of the roof.

A-9   Jamb sections are sectional views of windows and doors. A horizontal cutting plane splits open the window or door, exposing the internal components.

A-10   The three framing plans are floor, wall, and roof.

A-11   The detail stud plan shows a few studs at critical points, usually at intersections.

A-12   Electric heat does not require a duct system.

A-13   The different piping includes hot water supply and return lines to and from the heater boilers, natural gas or oil supply lines, refrigerant lines, condensate lines, as well as other kinds of piping.

A-14   Toilets, tubs, sinks, vents, traps, and sewer lines are normally shown on plumbing blueprints.

A-15   Dotted lines connect each fixture to its specific switch in a blueprint. These dotted lines are not the paths of the actual wires.

CHAPTER 8

A-1   The construction specifications include the types of materials that are used for construction, the required construction methods, the installation requirements for the building, the contractor's time frame.

A-2   The contractor should be familiar with the specifications, because they are a legal part of the contract. They also ensure no misunderstandings occur.

A-3   Same or equal quality means that a substitution of material is permitted, but the substitution must meet the standards of the original material.

A-4   The general information section includes the materials purchased, the work required, the estimated completion date, all drawings and legal documents.

A-5   The carpentry section includes the types of wood grades required, nail sizes, and the maximum amount of moisture allowed in the wood.

A-6   The answer may include three of the following: the quality of the plaster and lath, the type of workmanship to be used with the plaster and lath, the instructions for mixing and applying the plaster and lath, the drying time, number of coats, the acoustical requirements, floor treatments, and wall coverings (i.e., drywall).

A-7   The electrical section includes the electrical outlets, number of circuits, location of telephones, size of wire used, and a detailed list of all electrical components used (including style, manufacturer and model number)

A-8   Guarantees ensure the contractor uses the proper, quality materials and that the contractor installs each item correctly.

CHAPTER 9

A-1   A schedule contains information that may not be found on the construction specifications, such as the specific styles or colors of components.

A-2   Schedules are necessary to ensure no misunderstandings occur between the architect, engineers, and contractors. These misunderstandings include specific types of materials, styles, or colors.

A-3   Reference numbers show the location of the item on the blueprint.

A-4   Architects and engineers compile the schedules.

A-5   The material schedule is a list of the approximate quantities of all the materials that are to be used in the construction.

A-6   The contractor can use the material schedule as a checklist to ensure all the finishing touches in a particular building match.

A-7   MBH stands for one thousand Btu per hour.

ASSIGNMENT 1

A-1

	Kitchen	Dining Room	Family Room	Hall Bath	Master Bath	Bedroom 1	Bedroom 2
Number of supply registers	1	1	2	1	1	1	1
Size of each	14" x 6"	12" x 8"	10" x 8"	8" x 4"	10" x 6"	10" x 6"	12" x 8"
Model number for each	A611	A611	A611	A611	A611	A611	A611
Cfm to be delivered by each	200 cfm	150 cfm	125 cfm	50 cfm	100 cfm	100 cfm	150 cfm
Total cfm for each room	200 cfm	150 cfm	250 cfm	50 cfm	100 cfm	100 cfm	150 cfm

A-2   The return register is located in the hall.

A-3   The return register is Model 673 16" x 14".

A-4   There are 3 sizes: 6", 7", 8".

A-5   The register size is listed first, followed by the size of the round duct: 10" x 6"- 6"; 10" x 8"- 6"; 12" x 8"- 7"; 14" x 6"- 8".

A-6   The a/c condenser is located on the west side near outside deck.

A-7   The condenser model is UAFD-030JA and its Btu size is 30,000.

A-8   The furnace model is UGEB10EEGS and its Btu size is 100,000.

A-9   The evaporator model is UCEH-Z031 and its Btu size is 30,000.

A-10   The circuit breaker for the condenser is 40 AMP with #10 wire with ground.

A-11   The circuit breaker for the furnace is 20 AMP with #14 wire with ground.

ASSIGNMENT 2

A-1   Six areas in the building require cooling.

A-2   The cooled areas include the main office, reception area, bathrooms, office 1, office 2, lounge.

A-3   Eight areas in the building require heating.

A-4   The heated areas include the entrance, reception, bathrooms, rear hall, office 1, office 2, main office, lounge.

A-5

	Number of registers	Size of registers
Reception Area	1	12" x 12"
Bathrooms	2	8" x 8"
Main Office	1	12" x 12"
Office 1	2	10" x 10"
Office 2	2	10" x 10"
Lounge	2	10" x 10"

A-6

	Number of radiators	Size of radiators	Btu capacity
Reception Area	1	7'	4,200 Btu
Bathrooms	2	5'	3,000 Btu each
Main Office	2	5'	3,000 Btu
		8'	4,800 Btu
Office 1	2	5'	3,000 Btu
		8'	4,800 Btu
Office 2	1	8'	4,800 Btu
Lounge	2	8'	4,800 Btu each
Entrance	2	5'	3,000 Btu each
Rear Hall	1	5'	3,000 Btu

A-7   There are two ceiling mounted return registers.

A-8   The model number of the return registers is RETURN RHF45FILTER and each one is 24" x 48".

A-9   The total cfm for the building is 2000 cfm.

A-10   The duct sizes used are: 40"x8" - 16"x8" - 10"x8" - 8"x8" - 8" rnd - 6" rnd - 14" rnd.

A-11   The condenser is located on the north side of building near east end.

A-12   The central air conditioning unit is mounted in attic.

A-13   The boiler is located in the northeast corner of basement.

   The size, model number and manufacturer are: 100,000 Btu, GA92-100EI and Vaillant.

A-14   The manufacturer of the baseboard units is Heatrim Association.

A-15   The manufacturer of the condenser is Bryant Manufacturing.

ASSIGNMENT 3

A-1   For the supply lines, 4" pipe is used.

A-2   For the return lines, 4" pipe is used.

A-3   a)-   Gate Valve   ⊲⊳⊢

       b)-   Reducer   ─▷─

       c)-   Elbow 90°   ⌐

       d)-   Pipe Outlet Down   ──○

       e)-   T-Fitting   ⊥

A-4   Excluding the concrete pads, each boiler is 5' tall.

A-5   Fresh air enters the mechanical room through an intake louver.

A-6   Each boiler is approximately 4' 3" wide.

A-7   The boilers will fit through the door, as the door is 6' wide.

A-8   20" and 14" round flue pipes are needed.

A-9   The boilers are 5 feet apart.

ANSWERS FOR ASSIGNMENT 4

A-1   The condensate line has a diameter of 3/4".

A-2   When the condensate line empties into the drain, it has a diameter of 1 inch.

A-3   Each unit supplies 3 registers and each register supplies 665 cfm.

A-4   Flexible 14" round duct is used to connect the ceiling registers to the supply duct.

A-5   There is no return duct system connected to these systems.

HVAC AND BUILDING TERMS

A COIL: An air conditioning evaporator coil shaped like the letter A.

AGGREGATE: Materials mixed with cement to make concrete.

APRON: Wood trim located below the window sill; also a concrete pad or ramp.

AREAWAY: An opening or space below ground level used to permit light to enter a building.

ASPHALT: A black cement-like material used to waterproof outside surfaces, sometimes used on outside duct systems for waterproofing.

BACKFILL: Earth placed around foundation walls for filling and grading purposes.

BATTEN: A narrow strip of wood used to cover a joint. Also used to fasten two or more boards together.

BEAD: A molding used to protect an inside or outside corner.

BEAM: A horizontal structural supporting member.

BEARING WALL: A wall which supports weight other than its own.

BENCH MARK: A known point of reference from which other measurements are made.

BRANCH CIRCUIT: The portion of an electrical wiring system which stretches beyond the final overcurrent device.

CANTILEVER: A projecting beam that is only supported at one end.

CASEMENT WINDOW: A hinged window, usually with a metal frame, which opens out.

CHASE: Vertical or horizontal space left within a masonry wall for pipe, wire or ducts.

CIRCUIT: The electrical path which connects the source to the load and back to the source.

CIRCUIT BREAKER: A protective device that opens or closes the electrical circuit in order to prevent a damaging overload.

CLEANOUT: An opening at the lowest end of a sewer pipe or flue pipe to assist in the removal of foreign material.

COLUMN: A vertical support.

COMPRESSOR: A machine used in refrigeration systems to compress low-temperature, low-pressure refrigerant vapor into high-pressure, high-temperature refrigerant vapor.

CONDENSATE: The liquid that is formed when a vapor condenses.

CONDENSER: The device used to change or condense the refrigerant vapor into a liquid.

CONDUCTOR: A material, usually a wire, which carries electrical current.

CONDUIT: The covering used to carry the conductor, usually a metal tube or pipe.

CONVECTION: Transmission of heat by way of air current.

CONVECTOR: A surface which transfers its heat to a surrounding fluid by way of convection.

CRAWL SPACE: A small area between the underside of the floor and the ground.

CURRENT: The measurement of the flow of electricity in a circuit.

DIFFUSER: An air outlet which scatters air in various directions in order to promote the mixing of supply air with air already in a room.

DOUBLE-HUNG WINDOW: Window consisting of two sections that slide vertically up and down.

DOUBLE-STRENGTH GLASS: Glass which has a thickness of 1/8 inch.

DRYWALL: Interior wall material made of plaster board.

DUCT: A conductor made of sheet metal, used for distributing warm or cold air.

ELEVATION: The drawings showing the exterior or interior front, sides, and/or rear walls of a building.

ELBOW: A pipe or conduit fitting which has a bend, usually 90 degrees.

FIRE STOP: A wood block placed between the wall studs to prevent fire from spreading through the wall.

FLUE: A passage in a chimney which allows products of combustion to leave a building.

FOOTING: The base of a column or wall which distributes the weight of the building into the ground.

FRAMING: Wood or metal internal structure or building skeleton.

FROST LINE: The level below the earth surface to which frost can penetrate.

GALVANIZED: A zinc plating applied to steel to prevent rusting.

GAUGE: Usually refers to the thickness of metal or the diameter of wire.

GLAZING: Installing window glass in a frame.

GRADE: The ground level around a building.

GROUND: The connection between an electrical device and the earth.

HEADER: A supporting horizontal structure located above openings (i.e., doors and windows) which spreads the weight distribution from other structures.

HOSE BIB: A threaded faucet to which a garden hose can be connected, also called sill cock.

I BEAM: Steel beam manufactured in the shape of the letter I.

JAMB: The vertical sides of the door or window opening.

JOIST: Horizontal wood or metal items used to support floors, ceilings, and roofs.

LATH: Metal lath is wire mesh, which is used as a base for plaster. Wood lath is a narrow wooden strip, which is attached to studs and used as a base for plaster.

LEADER: Vertical pipe which carries rain from the roof gutter to the ground.

LINTEL: A steel beam which spans the top of an opening in a wall which carries the weight of the wall above it.

LOCK SEAM: The folding of two sheet metal plates together into one piece.

LOUVER: Horizontal slats that cover an opening, permitting air circulation.

MILLWORK: Finished woodwork in a building (i.e., cabinets, molding, door frames, etc.).

MORTAR: A combination of sand, water and cement which hardens in place; used in masonry to connect bricks and stone to one another.

PARAPET: The portion of the outside wall that extends above the roof.

PARTITION: An interior wall that divides one room into two rooms.

PITCH: The slope of a roof.

PLENUM: The chamber in a heating or air conditioning system into which air is forced and subsequently distributed to ducts.

PLUMB: Term used to describe an object that is in its true vertical position.

RACEWAY: A channel which contains the electrical wiring.

RAFTER: A structural member to which the roof is attached.

RECEPTACLE: An electrical outlet to which electrical appliances are connected.

REDUCER: A pipe coupling used for connecting pipes of different sizes.

REGISTER: A grille covering a duct outlet, which controls the velocity of the warm or cold air passing through it.

RISER: The vertical surface of a stair step.

SERVICE ENTRANCE CONDUCTOR: The electrical conductors which connect the building to the utility company supply point.

SETBACK: The minimum distance between the building and the property line.

SHEATHING: The first layer of material, usually plywood or wallboards, used to cover the outside wall studs.

SIDING: The finished material attached to the sheathing.

SQUARE: Refers to 100 square feet of material.

STACK: A vertical pipe used for venting purposes.

STORY: The space between a floor and ceiling.

STUD: The vertical posts which support the walls of a building.

THERMOSTAT: A heat-sensitive instrument that responds to temperature changes.

TRANSFORMER: An electrical device used to convert electric voltage.

TREAD: The horizontal surface of the stair, it is the actual step.

TRUSS: A structural piece in the shape of a triangle, used for supporting the roof over long distances.

VENT: A pipe extending outside the building to permit air flow.

WEATHERSTRIP: A strip of material that is applied around exterior doors or windows in order to reduce heat loss.

INDEX

Author Stanley Aglow has had more than 15 years of experience as an installation and service supervisor and writes a regular electrical service column for the *Air Conditioning, Heating and Refrigeration News*. Today, Mr. Aglow is the teacher and director of HVAC/R Training at the Philadelphia Wireless Technical Institute.

Other Titles Offered by BNP

TO RECEIVE YOUR UP-TO-DATE CATALOG,
CALL TOLL FREE
1-800-837-1037

**BUSINESS NEWS
PUBLISHING COMPANY**
Troy, Michigan
USA